BIO-INTELLIGENCE SCIENCE

Bio-Intelligence Science

Edited by:
Zoran Gacovski

ARCLER PRESS

www.arclerpress.com

Bio-Intelligence Science

Zoran Gacovski

Arcler Press
2010 Winston Park Drive,
2nd Floor
Oakville, ON L6H 5R7
Canada
www.arclerpress.com
Tel: 001-289-291-7705
001-905-616-2116
Fax: 001-289-291-7601
Email: orders@arclereducation.com

ISBN: 978-1-77361-079-5 (Hardcover)

Declaration

About the Editor

Dr. Zoran Gacovski has earned his PhD degree at Faculty of Electrical engineering, Skopje. His research interests include Intelligent systems and Software engineering, fuzzy systems, graphical models (Petri, Neural and Bayesian networks), and IT security. He has published over 50 journal and conference papers, and he has been reviewer of renowned Journals. In his career he was awarded by Fulbright postdoctoral fellowship (2002) for research stay at Rutgers University, USA. He has also earned best-paper award at the Baltic Olympiad for Automation control (2002), US NSF grant for conducting a specific research in the field of human-computer interaction at Rutgers University, USA (2003), and DAAD grant for research stay at University of Bremen, Germany (2008). Currently, he is a professor in Computer Engineering at European University, Skopje, Macedonia.

Contents

List of Contributors

Dr. Zoran Gacovski

Claudio Messori
Str. Villaggio Prinzera 1, Fraz. Boschi di Bardone, Terenzo, Italy

Juyang Weng
Department of Computer Science and Engineering, Michigan State University, East Lansing, MI, USA
School of Computer Science and Engineering, Fudan University, Shanghai, China

Tielin Zhang
Institute of Automation, Chinese Academy of Sciences, Beijing, China
University of Chinese Academy of Sciences, Beijing, China

Yi Zeng
Institute of Automation, Chinese Academy of Sciences, Beijing, China

Bo Xu
University of Chinese Academy of Sciences, Beijing, China

Necmettin Firat Ozkan
Department of Industrial Engineering, Eskisehir Osmangazi University, Eskisehir, Turkey

Emin Kahya
Department of Industrial Engineering, Eskisehir Osmangazi University, Eskisehir, Turkey

Chenxia Gu
Department of Mathematical Sciences, Xi'an Jiaotong-Liverpool University, Suzhou, China

Shaotong Wang
Department of Mathematical Sciences, Xi'an Jiaotong-Liverpool University, Suzhou, China

Hao Yu
Department of Mathematical Sciences, Xi'an Jiaotong-Liverpool University, Suzhou, China

Narusci S. Bastos
Federal University of Rio Grande (FURG), Rio Grande, RS, Brazil

Diana F. Adamatti,
Federal University of Rio Grande (FURG), Rio Grande, RS, Brazil

Cleo Z. Billa
Federal University of Rio Grande (FURG), Rio Grande, RS, Brazil

Romain Grandchamp
Laboratoire de Psychologie et NeuroCognition, Université de Grenoble, Grenoble, BSHM, 1251 av Centrale CS40700, 38058 Grenoble Cedex 9, France
CNRS, UMR 5105, Grenoble, France

Arnaud Delorme
Centre de Recherche Cerveau et Cognition (CerCo), Université Paul Sabatier, Pavillon Baudot, Hopital Purpan, BP 25202, 31052 Toulouse Cedex 3, France
CNRS, UMR 5549, Toulouse, France
Swartz Center for Computational Neuroscience, Institute of Neural Computation (INC), University of San Diego California, La Jolla, CA 92093-0559, USA

Terence Lovat
The University of Newcastle, Newcastle, Australia

Daniel Fleming
The Broken Bay Institute, Sydney, Australia

Hong Gi Yeom
Interdisciplinary Program in Neuroscience, Seoul National University, Seoul 151-742, Republic of Korea

June Sic Kim
Department of Brain and Cognitive Sciences, College of Natural Sciences, Seoul National University, Seoul 151-742, Republic of Korea

Chun Kee Chung
Interdisciplinary Program in Neuroscience, Seoul National University, Seoul 151-742, Republic of Korea
Department of Brain and Cognitive Sciences, College of Natural Sciences, Seoul National University, Seoul 151-742, Republic of Korea
Department of Neurosurgery, Seoul National University Hospital, Seoul 110-744, Republic of Korea

Kai Li
Electrical Geodesics, Inc., 500 E. 4th Avenue, Eugene, OR 97401, USA

Xenophon Papademetris
Electrical Geodesics, Inc., 500 E. 4th Avenue, Eugene, OR 97401, USA
Yale University, New Haven, CT 06520, USA

Don M. Tucker
Electrical Geodesics, Inc., 500 E. 4th Avenue, Eugene, OR 97401, USA

Jianping Li
School of Computer and Information Technology, Northeast Petroleum University, Daqing, China

Haizhou Wu
College of Information Science and Engineering, Guangxi University for Nationalities, Nanning 530006, China

Yongquan Zhou
College of Information Science and Engineering, Guangxi University for Nationalities, Nanning 530006, China
Key Laboratory of Guangxi High Schools Complex System and Computational Intelligence, Nanning 530006, China

Qifang Luo
College of Information Science and Engineering, Guangxi University for Nationalities, Nanning 530006, China
Key Laboratory of Guangxi High Schools Complex System and Computational Intelligence, Nanning 530006, China

Mohamed Abdel Basset
Faculty of Computers and Informatics, Zagazig University, Zagazig, Egypt

Preface

Today, it is considered that intelligence includes at least two skills: the ability to memorize and store knowledge, and the ability to process knowledge. The person (or machine) without any knowledge cannot be considered intelligent. The ability of learning - acquisition of new knowledge, is also one of the aspects of the intelligence, although we can classify it as an ability to solve problems. As an "intelligent feature" we can also consider the ability to communicate with other intelligent beings.

For the concept of intelligence - two questions are essential: the question of knowledge and the reasoning (making conclusions), and, this corresponds to the terms of a knowledge base and a reasoning process. The component of reasoning (inference) also represents a kind of knowledge - it is knowledge about the process of carrying out new information from an existing knowledge base.

This edition covers different topics from bio-intelligence science, and application of bio-intelligence in different domains - the bio-medical domain, the learning, the medicine etc.

Section 1 focuses on biological aspects of the intelligence, describing biological vs. artificial intelligence, brain as an emergent finite automaton, biological neural network structure and spike activity prediction based on multi-neuron spike train data, an experiment in use of brain computer interfaces for cognitive researches, and chessboard model of human brain and an application on memory capacity.

Section 2 focuses on topics from neuroscience, describing patterns discovery in brain signals using decision trees, an interactive immersive tool for brain education, art, and neuro-therapy, modeling neuromorphic persistent firing networks, and creativity as central to critical reasoning and the facilitative role of moral education.

Section 3 focuses on pattern recognition in neuro and medical applications, describing Brain-k for structural image processing: creating electrical models of the human head.

Section 4 focuses on neural networks applications, describing quantum-inspired neural networks with application, training feedforward neural networks using symbiotic organisms search algorithm.

Editor

INTRODUCTION

Dr. Zoran Gacovski

BACKGROUND

Under intelligence - we normally mean the ability to acquire, memorize and process certain knowledge. However, no definition of biological intelligence is generally accepted (at least some corresponding to the common, everyday sense, and at the same time - that precisely describes the term). In any case, we can consider that intelligence includes at least two skills: the ability to memorize and store knowledge, and the ability to process knowledge. The person (or machine) without any knowledge cannot be considered intelligent. Also, as intelligent cannot be considered a person (or machine) with huge amounts of "static" knowledge or information, but without the ability to handle that knowledge to solve problems related with it. There are other aspects of intelligence. One of them is the speed of knowledge processing, which in case of a machine, depends not only by the algorithm, but by the hardware (brain) as well.

The ability of learning - acquisition of new knowledge, is also one of the aspects of the intelligence, although we can classify it as an ability to solve problems. As an "intelligent feature" we can also consider the ability to communicate with other intelligent beings (machines), which can also be treated as problem solving. Therefore, we will consider that being or machine to have intelligent attributes, if it has the above properties. If it is a machine or a device with the specified properties, we will consider it as a bio-inspired artificial intelligence.

For the concept of intelligence - two questions are essential: the question of knowledge and the reasoning (making conclusions), and, when it comes to computers, it corresponds to the terms of a knowledge base and a reasoning component. The component of reasoning (inference) also represents a kind of knowledge - it is knowledge about the process of carrying out new information from an existing knowledge base and the rules by which this performance takes place. Knowledge of this type is called meta-knowledge. The original concept of knowledge also has more aspects. According to many philosophical concepts - the knowledge represents "statements" for which it is determined, or can be determined - to be truth.

Neural networks represent one of the main tools in the artificial intelligence. Instead of conceptualizing the problem in mathematical form, neural networks use the principles of the human brain and its structures, to develop a data processing strategy. In this section we will explain some of the main concepts and core principles related to neural networks. Since neural networks have developed on the model of the human brain, we will expose the biological aspects for the development of the neural networks.

The development of the neural networks has happened not far into the past (they were introduced in the 1940-s). Today, the methods they provide - are widely used, both in the technical, and in social sciences, and some of the most relevant applications are in the IT, medicine, economics etc. Applications in the economy are mainly linked to the prediction of possible future values, which can relate to predicting of sales volume depending on the marketing activities undertaken, predicting stock prices on a certain date (important for banks and investors), and the like. A precise prediction of the values of the relevant variables is essential for the operation of all business entities. In addition to the use of neural networks for the purpose of forecasting, another significant usage is for the purpose of classification. Neural networks great performance in classification and prediction - qualifies them as excellent means for multivariate analysis.

BIO-INSPIRED NEURAL NETWORKS

The development of neural networks originates from 1943 - then Warren McCulloch and Walter Pitts, introduced in their work the first simple mathematical model of a neuron, as the basic unit of the nervous system. This model was discrete and based on binary logic. They showed that certain logical and arithmetic functions can be calculated, even with simple types of neural networks. However, the lack of their model was that it was not capable of learning. On the other hand, their work has had an impact on other researchers. Thus, the founder of cybernetics, Norbert Wiener, was inspired by this model when it examined the similarities between the nervous and the computer system. Next to him, the author of the American project of the electronic computer, John von Neumann, in several of his works proposed construction of computers that are based on the functioning of the human brain.

In 1949 Donald Hebb wrote a book titled "The organization of behavior", in which he described the paradigm of learning, which bears his name (Hebb-s doctrine), which is based on the view that the permeability of the synapses (neuron's joints) increases if the activities of two adjacent neurons are in conjunction for a certain time. This principle is essentially tied to update of the links. The book had considerable influence on scientists in the 40s and 50s of the twentieth century, which have further developed their views. The main representative of this period is Minsky Marvin, who is credited with the first application of neural networks. In 1951, Minsky together with Dean Edmonds designed the first neurocomputer Snark (40 neurons), with synapses that have capability of calculating the values of the weighting factors (extent of synaptic permeability), depending on the performance of a specific task execution (which is based on Hebb-s mode of learning).

In 1957 - Frank Rosenblatt invented the perceptron, which is based on the McCulloch-Pitts neuron's, which has a learning algorithm, i.e. a way to change the values of weight coefficients by iterative method, in order to obtain a desired result. Shortly after the discovery of perceptron, Bernard Widrow and his student Marcian Hoff, have developed another type of neural computational element, which is called ADALINE (Adaptive Linear Element), and which included a new efficient learning rule, which was applied for adaptive signal processing, in control systems etc.

This expansive development of neural networks abruptly stopped by publishing of the extremely influential books "Perceptron", which was written by Minsky and Papert, and published in 1969. It expressed great pessimism when it comes to the possibilities for a single-layer perceptron, stating that it cannot be used to solve some simple logical functions. On the other hand, they recommend that more attention should be directed to the development of knowledge - based, i.e. expert systems.

In the mid-eighties neural networks have experienced a renaissance, thanks to the sharp rise in researchers interest in this field. Amari in Japan has studied the mechanisms for training of neural networks and their mathematical background. The Japanese researcher Fukushima has developed a new class of neural networks, known as Cognitroni and Neocognitroni. Substantial research in the field of associative memory was performed by Kohonen in Finland. One of his most significant discoveries is the learning without supervision. Great discoveries in the field of neural networks were made by Carpenter and Grossberg, which are attributed for the development of ART networks (Adaptive Resonance Theory). The architecture-recurrent neural network for associative memory was firstly developed by Hopfield (in the 1980-s).

NEURAL NETWORK'S STRUCTURE

The human brain and the entire nervous system are made up of nerve cells - neurons, which are about 10^11, and they can have around 10^15 connections between them. The neuron has four main parts:

- Cell's input part,
- Cell's body (soma),
- Cell's output part, and
- Synapses.

The input part of the cell contains a set of branches - called dendrites. The body of the cell processes the signals received from the dendrites, thus obtaining the output pulse that is sent to all ends of the branches - called an axon, which is the output part of the cell. The place where the axon touches the dendrites of another cell is called a synapse. This is where impulses are transmitted from one to another nerve cell (called, cellular transduction).

In analogy with the biological neuron, a model can be formed for the artificial neurons. For representation of the artificial neurons, we will introduce the following variables:

- x_i (i = 1, 2, ..., m) - input connection, the value of inputs that neuron receives (signals, or input potentials in the biological neurons);
- w_{ki} (i = 1, 2, ..., m) - the so-called weights (synapses in biological neurons);
- u_k - the aggregate value, which is the sum of weighted inputs (body of the biological neurons);
- a (.) - activation function (axon in biological neurons);
- Threshold;
- y_i - the value of the output (output potential in the biological neurons).

Schematic diagram of such neural network is presented in the figure 1. Each of the input connections, which are marked with x, (and which are m in number), is multiplied by its weight coefficient w_{ki}. These multiplied values are then added together, and so we obtain the aggregate value net = u_k. So, we have:

$$u_k = net = \sum_{i=1}^{m} w_k x_i$$

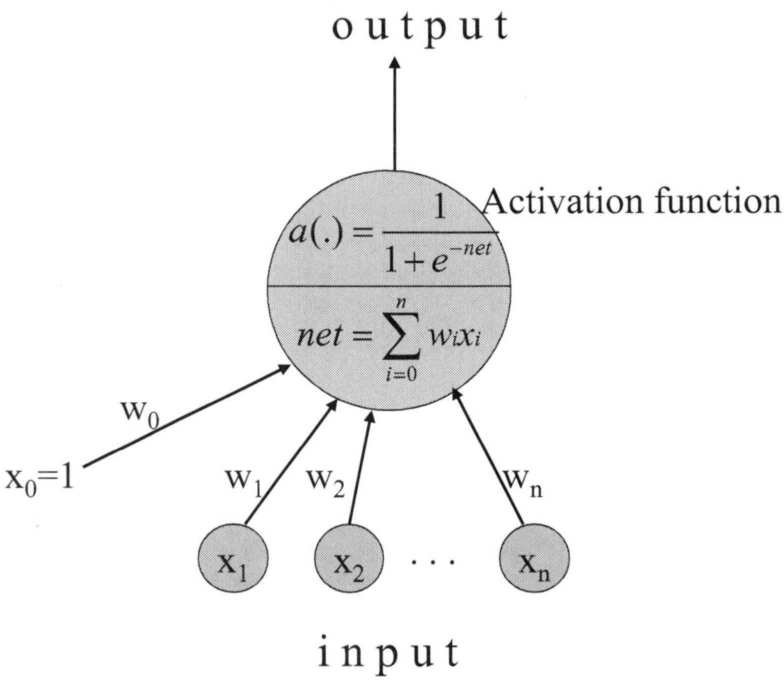

Figure. 1. Schematic diagram of an artificial neuron.

Then, the value u_k is used as an input in the activation function, which is generally non-linear, and usually it is the so-called sigmoid (logistic) function (Fig. 2). The link between the thresholds θ_k and the activation function is usually such that the value θ_k is subtracted from the value of the u_k, so, this difference is used as input to the activation function. Thus - the output value is calculated, i.e. the output y_i.

The topological architecture of the neural network represents a specific arrangement and connectivity of its neurons in the grid. By architecture, neural networks vary according to the number of neural layers. The best-known neural network with one layer is a single perceptron. In the networks with multiple layers, each layer typically receives input from the previous layer, and the outputs are sent to the next layer.

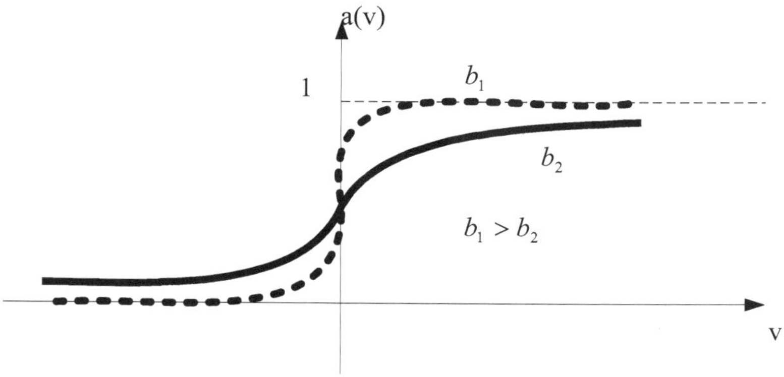

Figure. 2. Sigmoid logistic - activation function (threshold).

The first layer is referred to as input, the last as output, the inner layers are usually called hidden layers. One of the most common neural network architecture is a network with three layers. The first layer (input) is the only layer that receives signals from the environment. The first layer transmits the signals to the next layer (hidden layer) that processes the data and extracts the features and patterns from the received signal. The data that are considered important are forwarded to the output layer, last layer of the network. On the output of the third layer - the final results are obtained (calculated). More sophisticated neural networks may have more hidden layers, feedback loops and time lag elements, which are designed to enable more efficient extraction of the essential properties from the input layer. Today, multi-layer neural networks are mainly studied and applied, and in addition to the input and output layers they have neurons in the middle (hidden) layers. The Fig. 3 represents a neural network with three layers.

According to the direction of information propagation through the neural networks - they can be divided into:

- Feedforward (no-recursive, non-recurrent, or irreversible) - Higher layers do not return information to the lower layers. They propagate signals in one direction only (from the inlet to the outlet). A representative of this type of neural networks is the multilayer perceptron, or network with multiple layers.
- Feedback (recursive or recurrent or reverse) - Higher layers return information back to the lower layers. The output of

the neurons is returned to lower layers, or to the same layer. Representatives of these types of networks are: Hopfield, Cellular Neural Network (CNN), Kohonen neural network, double-associative network, etc. Feedback networks have much greater processing capability than the feedforward networks.

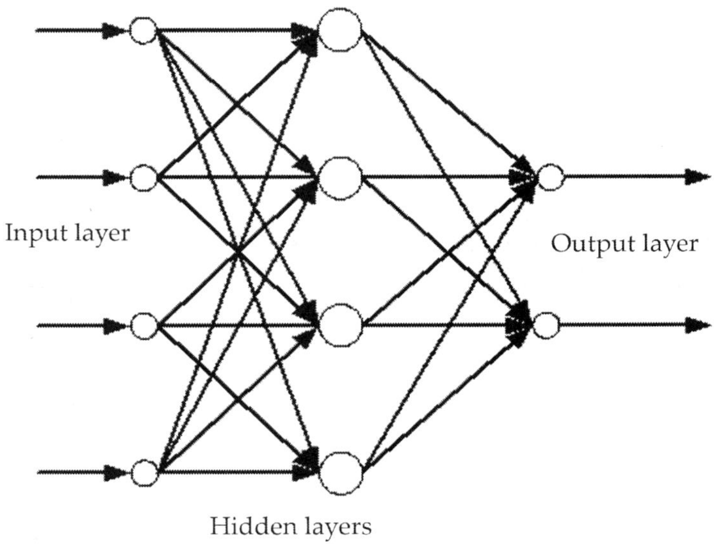

Fig. 3. Diagram of a neural network with three layers.

An important feature of neural networks, which is substantially separated from other multivariate techniques, is their ability to learn. According to the type of learning in the network, neural network algorithms are divided into:

1. Supervised learning - the values of the output variables in the network are known (Backpropagation, Radial Basis function, Probabilistic network), and
2. Unsupervised learning - the values of the output variables in the network are not known (Principal component, ART networks).

In particular, there are a few examples of test problems using these two algorithms:

- Prediction of a company bankruptcy, with the knowledge that the company had similar problems in the past (or not);

- Credit rating of a client, with knowledge that there was credit in the past, repaid on time, or not;
- The behavior of customers in relation to a product with a known customers' past actions, or with unknown past actions.

It should be noted that supervised learning is mostly used. In this case, the current values are compared with the output values. If there is a difference between these two values, we can access so-called coaching system. The most common form of training is the backpropagation algorithm, in which the error in the output value is calculated and sent back through the system of neurons, where there is a proportional change in the value of weight coefficients, which can increase or decrease, depending on the direction of the error. When these values are changed, inputs from the second layer enter the system and the process is repeated. The goal is to process a large number of cases through the neural network in the training phase, in order to ensure the highest quality output compared to existing inputs.

The flexibility of neural networks enables their application in a wide variety of problems, ranging from forecasting to the classification, time series analysis etc. However, the applicability depends on the specifics of the problem. Problems in which neural networks can successfully be applied are those in which:

1. There are many previous examples / representatives (at least about 100, and preferably more);
2. Rules for a decision making are not known (black box), i.e. experts are not available;
3. Variables can be quantified (which in most cases is possible);
4. Standard statistical methods showed no success, i.e. the phenomenon cannot be presented by some linear model;
5. Behavior phenomena are uncertain, data is incomplete, and a robust tool is needed.

REFERENCES

1. Journal of Computational Intelligence and Neuroscience, Hindawi Publishing, vol. 2017, 2017. doi: 10.1155/8483/, ISSN: 1687-5273.

2. Journal of Neuroscience and Medicine, *Scientific Research publisher*, vol. 7 no. 1, 2016. doi: 10.4236/nm.2016, ISSN: 2158-2947.

SECTION 1: BIOLOGICAL ASPECTS OF THE INTELLIGENCE

CHAPTER 1

INTELLIGENCE VS. ARTIFICIAL INTELLIGENCE: THE KING IS NAKED

Claudio Messori

Str. Villaggio Prinzera 1, Fraz. Boschi di Bardone, Terenzo, Italy

ABSTRACT

The sociological dimension of Artificial Intelligence (AI), and automation, is placed in the path traced by the man-machine integration process, started in the eighteenth century by the Industrial Revolution that assumes its current connotation after World War II. The use of the term intelligence, which appears in the expression Artificial Intelligence, is shown to be improper and misleading, and the expression itself should be replaced, eventually, by Artificial Simulation of the Savant Syndrome. By adopting a theoretical perspective, namely Endo-Dynamo-Tensive Model [Messori 2012B], it is traced the mapping of human's neurological (neuro-dynamics) and psychological (psycho-dynamics) dimension, and is provided the coordinates to be followed in the phenomenological qualification and definition of the psychic-mental- cognitive function intelligence. The neuro-dynamics of the Nervous System (NS), is taken into account by adopting the theoretical, interpretative and investigative perspective indicated by the particular line of research developed in the context of Quantum Field Theory (QFT) and QED (Quantum Electrodynamic Field Theory), that describes the water of which all living systems are composed, i.e. biological water, as water in a coherent oscillatory phase or state other than that of common water, named super-coherent oscillatory state. By introducing the possible functional role exerted within brain activity by glial cells, cerebrospinal fluid, intra- and extra- cellular fluid is outlined the overcoming of the classical neuroscience paradigm, based on the vision of brain activity as ruled by networks of neurons interconnected by synapses. The body-mind hard problem is taken into account and a solution is advanced. The psycho-dynamics of the humans mind territory is taken into account according to the four poles of mental functions introduced by C.G. Jung in his Psychological Types (1921), where he introduces a hierarchy of mental functions in two mental bipolar dimensions (dichotomies). These are sensing (attentiveness by means of the sense organs) coupled to intuition (awareness in unconscious way or being aware of unconscious contents) and thinking (function of intellectual cognition; the forming of logical conclusions) coupled to feeling (function of subjective estimation). At the conclusion of this work, it is provided the phenomenological definition of intelligence that does not contemplate the possibility to apply to neuroscience, and to natural and human sciences in general, the paradigm that inspires the research on AI, i.e. computational model and Information Theory.

Keywords: Artificial Intelligence, Man-Machine Integration Process, Artificial Simulation of the Savant Syndrome, Endo-Dynamo-Tensive Model, Biological Water, Autopoietic Psycho-Neuro Dynamics, Intelligence

ARTIFICIAL INTELLIGENCE: FALSIFYING THE IDENTITY OF THE MENTAL PHENOMENON AND ITS FUNCTIONS

Groucho Marx said that "Military intelligence is a contradiction in terms". Well, AI is the hyper-technological product of this contradiction, and I guess Norbert Wiener was right in saying that "...wartime technology, which has furnished from small bits of scientific truth great warehouses of potential threat and arsenals of destruction, has much to thank scientist for their intellectual prostitution has been significant". Anonymous

To possess and exert control is the mantra of the third millennium, and the computerization of society meets this purpose: "The computerization of society...has essentially been a side effect of the computerization of war". (Frank Rose). The globalization of computerization is a project planned by the US military fields during the years of the Cold War[1], which offers to those who run it the possibility to control almost everything based on telecommunications and on computer processes. The Artificial Intelligence (AI) is strategically located in this project.

Artificial Intelligence, a term coined in 1956 by the young American mathematician John McCarthy, ranks in the footsteps of the man-machine integration process started by the eighteenth-century Industrial Revolution, and is the legitimate daughter of the Information Theory, a discipline born in the Telecommunications field (telegraph and telephone) between the 20's and 30's and developed between the 40's and 50's separately by Claude Shannon (The Mathematical Theory of Communication) and Norbert Wiener (Cybernetics). The term intelligence that appears in the expression coined in 1956 by McCarthy is a legacy of the terminology used by one of the pioneers of Information Theory, Harry Nyquist of the American Telephone and Telegraph Company (AT & T). In 1924 Nyquist publishes an article on the Bell System Technical Journal (BSTJ) entitled Certain Factors Affecting Telegraph Speed, where deals with the factors that affect the "maximum speed of transmission of intelligence". For those in the works of the period, the metaphorical term "intelligence" used by Nyquist for the transmission of an electromagnetic signal, appeared improper and misleading, compromised by anthropomorphic and psychological references incompatible with the subject matter. The transmission of signals between machines (coder/decoder), i.e.

the sending and receiving of electromagnetic state variations through a medium (broadcast or via cable), could not be in any way confused with the transmission of meaning (messages) and neither could be associated with human properties generated by the possibility to feel, to sense, to think, to perceive, to intuit, to hypothesize, to imagine, to invent, to reflect, to dream, to experience. Four years after the publication of Nyquist's article, a colleague of Bell Telephone Laboratories Inc., Ralph V.L. Hartley, published an article on the BSTJ, entitled Transmission of Information, where the intelligence metaphor is replaced, for reasons of "physical as contrasted with psychological considerations", by the metaphor information.

But what did Nyquist mean by the term intelligence? In fact nothing would have to deal with human properties as those listed above, or could attribute to a signal a semantic meaning. By the term intelligence, Nyquist refers to the statistically determined and decipherable component of a random signal (i.e. to the data inputs, conveyed and made available by an analogic signal, qualified by Nyquist as "the number of characters, representing different letters, figures, etc." transmitted in a certain period of time), whose transmissibility (with no loss of intelligence i.e. data) from a transmitting device (coder) to a receiver (decoder) depends on the degree of uncertainty resolution (associated with the noise inherently generated by the transmission medium or by the involved apparatuses) obtained in the signal transmission.

But times change, and what in the 20's for reasons of "physical as contrasted with psychological considerations" could not be described as intelligent, it became in the 50's.

From the 50's onwards, a wide array of scientists, researchers, scholars and experts, who for various reasons and to varying degrees have conformed to AI, have started adopting the computational model on which AI is based, transferring it to biology, genetics, psychology, psychiatry, anthropology, ethology, neuroscience, so as to make the mental phenomenon (the res cogitans of Cartesian memory) somehow objective (according to the Cartesian dichotomy res cogitans is a fictitious, not objective entity, and as such of no interest to science), a phenomenon that ever since is governed by the same laws that regulate the transmission of signals fitted with meaning in telecommunications (coder and decoder exchange messages → cells also exchange messages) and in computer programming (computer coding is based on computation → genetic coding also is based on computation) is no longer as fictitious as it seemed. In doing so, the in vivo reality has been forged in the image and likeness of the in vitro Information Theory.

Unconcerned of the fact that ".the brain is not a bundle of neurons dissected in a laboratory but exists within an organism which is essentially carrying on its own auto-regulation, its nutrition and self-preservation, which feels hunger and thirst and needs social relations, while continually permeated, interwoven, by emotions, sentiments, needs, desires"[2], these AI followers insist on locating the brain functions in this or that area and neuronal network, in this or that sub-cellular (e.g. microtubules) or inter-cellular (e.g. synapses) structure, in this or that sequence (code), deterministic or probabilistic, of information, and not pleased with that come to resurrect the adult- metric vision of some pseudo-scientific tradition of the past, according to which ".the human infant is a decerebrated being equipped with reflexes". (Paul Emil Flechsig, 1847-1929), and "The Negro brain possesses a spinal cord of the type found in children and women, and beyond this, approaches the type of brain found in higher apes" (Emil Huschke, 1797-1858), to discredit the cognitive level of a child, associating his IQ not anymore to that of a de-cerebrated being or to that of a higher ape, but to that of a computer, alias intelligent machine.

That's what has been done in 2013, for example, by an AI experts team of the University of Illinois at Chicago (UIC), which has designed and conducted a study[3] the purpose of which was to evaluate the degree of "intelligence" of the super computer ConceptNet 4, by administering to the machine a part of a well known IQ test, administered as IQ test of children aged between two years and six months and seven years and seven months, the Wechsler Preschool and Primary Scale of Intelligence Test (WPPSI) created in 1967 by American psychologist David Wechsler, already creator of the Wechsler Intelligence Scale for Children (WISC, 1949) and of the Wechsler Adult Intelligence Scale (WAIS, 1955). The study suggested that ConceptNet 4 would have an IQ comparable to that of a four-year age child.

Studies like this demonstrate two things, the first is that even in the twenty-first century kids are being targeted by new detractors of their intelligence, and the second is that the temptation to falsify the physical identity of the mental phenomenon, and its functions including intelligence, trying to pass the camel intelligence through the eye of the needle of telecommunications, information technology and cybernetics is too high so as not to bear fruits.

The definition of intelligence provided by David Wechsler goes exactly in this direction: Intelligence is the aggregate or global capacity of the individual to act purposefully, to think rationally and to deal effectively with his environment (Wechsler, 1944).

Act intentionally, think rationally, effectively deal with the environment: are these the conditions that qualify intelligence and make a being intelligent?

Intentionality and the use of rational thinking for adaptive and supra-adaptive purposes, are very late events in the history of mankind [1] . The occurrence of a behavior clearly moved by intentionality dates back to the Middle Paleolithic, while for rational thinking strictu sensu we must wait for the Upper Paleolithic: does it means that before that, human communities were deprived of intelligence? The relational life of a two year old child is governed by the symbolic way of thinking and if left to himself will hardly be able to survive: can we consider him or her intelligent? The magical thinking that characterizes the social life of an indigenous belonging to a settled community of the Amazon forests, which for generations has never abandoned his ancestral territory, and has never changed his own traditions and his own means of livelihood, makes him an intelligent being? ConceptNet 4 can really be considered somehow intelligent?

Claim to measure intelligence with a test means to apply to the res cogitans the same reductionist and mechanistic pattern applied by the positivist paradigm to the res extensa: the structuring of the objects and phenomena that we observe inside and outside of us, happens thanks to a sequence of combinations (factorizable), guided by a code (algorithmic) among certain structural elements (building blocks). In the case of a material object, the structural elements can be molecules, atoms, particles. In the case of a fictitious object as intelligence can be, e.g. the logical-mathematical intelligence, the verbal i., the spatial i., the musical i., the kinesthesic i., the emotional i. (Howard Gardner), in turn decomposable into sub-groups, and so on and so forth. Once broken down into objectifiable structural elements, the object shall be subjected to measuring and eventually reproduced and controlled.

The result is a puzzle composed by many elements, connected to joint, in a frame which develops a given function. But in the in vivo reality things are very different, and it is not the structure that generates the function, while the opposite is true.

Artificial Intelligence: The Man-Machine Integration Process

If the seventeenth and early eighteenth centuries are the age of clocks, and the later eighteenth and the nineteenth centuries constitute the age of steam engines, the present time is the age of communication and control. Norbert Wiener

The profile provided today of the res cogitans, must be functional to the modern application of the teaching imparted to history by the eighteenth-century Industrial Revolution: thanks to the positivist machines, invented and realized with the ingenuosity and the hands of skilled artisans and the money of daring patrons of the eighteenth and nineteenth-century bourgeois urban class, open up perspectives for profit and wellbeing (but also of destruction and death, the other side of the coin) unthinkable without them, a true technological Eldorado. The entire socio-economic system of almost the entire planet would not exist without the introduction of mechanical systems in the production cycle. The very idea of the production cycle belongs to the man-ma- chine integration process, started in the eighteenth century in the workplaces, that following an entreprencurial logic puts the human capital (the workforce) in a position of inferiority and subordination to the technological capital (technicians, machines, weapons). When patrons became entrepreneurs and artisans technicians, science and scientists became an essential part of this man-machine integration process. AI belongs, in fact and in law, to it.

Over the lifetime of a septuagenarian born in the late nineteenth century, Western society was driven by an unshakable faith in the power of science and technology, and technology itself was transformed from a haphazard process of invention to a science- based discipline. Steam engines and railroads, electric lights and airplanes, the telegraph, typewriter, motion pictures, radio communication, radar systems and plastics were just some of the achievements reached by the Western scientific and technological miracle between late nineteenth and the mid-twentieth, when by conceiving, building and using the first weapons of mass destruction (e.g. rockets, atomic bomb, hydrogen bomb, intercontinental ballistic missiles), the miracle showed to history its terrible destructive power.

Between the 50's and 70's, in the shadow of the devastating effects produced by the launching of the atomic bomb on Hiroshima (Little Boy) and Nagasaki (Fat Man), the machines definitely come out of crafts and industrial productive perimeter to become television sets and household appliances, partners that bring advantages and comfort in the homes' habitat of millions of people, integral part of their lives and relationship dynamics, differentiating themselves according to the consumers tastes and expectations, to set up the conditions for what will become a gradual addiction to the availability of technological solutions, more and more disconnected from the real needs of their users and increasingly self-referential.

The years between the 60's and 90's mark a new qualitative leap in the man-machine integration process, a new technological and psychological revolution that thanks to the fortunate combination between personal computers, World Wilde Web and the scion AI, whose mission is "designing systems that exhibit the characteristics associated with human intelligence, like understanding language, learning, reasoning, solving pro- blems, and so on", and whose belief is that "every aspect of learning or any other feature of intelligence can in principle be so precisely described that a machine can be made to simulate it", transforms the machine in a metacognitive tool, in an interactive platform that can assist you in work, study, communication, socialization, in leisure, entertainment, in portal to have access to a virtual ocean where you can draw everything and its opposite. From now on, the machine is no longer just part of our habitat, the machine becomes part of us and we part of it. It is in this period of time that those who deal with AI mature the conviction that to perform the spell to animate the machine and make it as one of us, to make the illusion of its presumed intelligence the more likely, it needs to be equipped by the ability to speak, and this is one of the task of the Strategic Computing Program (SCP).

In the 90's the know-how gained during the 80's by computer science, cybernetics, robotics and AI research, contributed to the realization of a project, the SCP, financed with funds both public (from the omnipresent

American federal agency Defense Advanced Research Projects Agency, DARPA) and private (from investors as the industrial giant IBM, the Dragon Systems company, the BBN, i.e. Bolt Beranek and Newman, and the Systems Development Corporation, SDC), thanks to which will come out the so-called intelligent machines, namely: machine with advanced intelligence technology and high-performance computing, including speech recognition and understanding, natural-language computer interfaces, vision comprehension systems, and advanced expert systems development, provided by a significant increasing in computer performance, through parallel-computer architectures, software, and supporting micro- electronics.

Artificial Intelligence: Artificial Simulation of the Savant Syndrome Leading to Cyborgs

The emperor walked beneath the beautiful canopy in the procession, and all the people in the street and in their windows said, "Goodness, the emperor's new clothes are incomparable! What a beautiful train on his jacket. What a perfect fit!" No one wanted it to be noticed that he could see nothing, for then it would be said that he was unfit for his position or that he was stupid. None of the emperor's clothes had ever before received such praise. "But he doesn't have anything on!" said a small child. Hans Christian Andersen (The Emperor's New Clothes)

Drones remotely controlled via Advanced Drone Interfacing System. Anthropomorphic robots interfaced in the Cloud Machine. Vocal assistants interconnected in the cloud environment supported by a neural net…

The dawn of the third millennium marks the spread on a global scale of very performing computing devices for home, medical, scientific, military, industrial use, that thanks to sophisticated voice recognition and production systems make them able to entertain, actively and in real time, an exchange of verbal information (although virtual, because you are anyway talking to a computing platform masquerade as human interlocutor), computationally itemized and statistically endowed with meaning. A fact equipped with enormous evocative implications (fascination).

In giving voice to a machine, reemerges the magic power assigned to the word-sound by the mid and late Paleolithic human communities [1] [2] [3] , subsequently merged in the post-Neolithic metaphysical and philosophical concept of psyche, pneuma, breath, ruah, prana, anima/animus, and finally in that of Energy (energheia), i.e. the ability attributed to the acoustic vibration to animate, to give life to the inanimate (a skill of which the male gender was soon made depositary, imposing the semantic baptism as an act of legitimation of reality, to be able to contend and to pull out from the female uterus her natural power to generate life).

The scientist who gives voice to a machine is like the shaman who gives voice to the spirits: the spirits speak through the shaman's intercession, machines speak through scientist intercession. The task of the Neolithic shaman (and after him of any other person appointed to act as an intermediary, medium, with the otherness) was to curry favor with the spirits, so that they could influence the forces of Nature, for the good (or for bad, black magic vs. white magic) of the community. The task of the contemporary shaman-scientist is much more ambitious: he wants literally subdue and even replace the forces of Nature. Meanwhile he succeeded, thanks to artificial insemination in vitro, to pull out from the female uterus (and not just thanks to the speech monopoly, but in deeds, artificially) the natural power of generating life belonging to woman.

Here, in this bacchanal that involves talking machines and a wide array of scientists, researchers, scholars and experts, who for various reasons and to varying degrees have conformed to the AI Eldorado, makes its way the idea of a future populated by intelligent machines equipped with consciousness (androids), perhaps in service to humanity or maybe not and maybe, as some claim (transhumanists), smarter, more reliable and better performing (Ray Kurzweil, chief engineer of Google, theorizes the overcoming of human intelligence by computers), definitely more programmable and controllable than humans.

If we'll have to compete with more and more "advanced generations of machines", somehow equipped with consciousness, a question arise: is there any kind of intelligence in AI, or to pursue the development of human-machine hybrids, imagining them as a first step towards the production of artificial humanoids, somehow better than humans, is a project that resuscitates eugenics in an AI format?

60 years have passed since John McCarthy coined the term Artificial

Intelligence and what we can say about the research in the AI field is that: is conducted by a range of scientists and technologists with varying perspectives, interests, and motivations. Scientists tend to be interested in understanding the underlying basis of intelligence and cog- nition, some with an emphasis on unraveling the mysteries of human thought and others examining intelligence more broadly. Engineering-oriented researchers, by contrast, are interested in building systems that behave intelligently. Some attempt to build systems using techniques analogous to those used by humans, whereas others apply a range of techniques adopted from fields such as information theory, electrical engineering, statistics, and pattern recognition. Those in the latter category often do not necessarily consider themselves AI researchers, but rather fall into a broader category of researchers interested in machine intelligence[4].

Is there any kind of intelligence in AI?

In the best case, the performances of the so-called intelligent machines are and will remain a computational, artificial simulation of some abilities manifested by people affected by the savant syndrome[5], or syndrome of the "idiot savant", where idiot is referred to a male (savant syndrome occurs approximately four times more frequently in males than females as does the incidence of autism) suffering from a series of more or less severe cognitive and mental disabilities, but that has one or more than one super- developed abilities, a jarring juxtaposition of ability and disability in the same person always combined with massive memory (typically eidetic memory). This is all that an intelligent machine can and will aspire to do: artificial simulation of the savant syndrome (ASSS). What the ASSS experts can do and will be able to do is to develop and get the most out of the super-memory (storing) and computational skills that qualify an intelligent machine, inventing new algorithmic systems, new software and new media devices for new applications (e.g., via emotion modelling [4] the machine will be programmed and tuned on the affective history, memories and aspirations, disappointments and expectations, desires and frustrations of their owner, so to satisfy at best their wishes and demands).

Realize an intelligent machine equipped with consciousness? Ridiculous. What can be done, and that is already being done, is to acquire new persuasive and pervasive technological tools, including robotic and computer systems exploiting AI, to be used in the military and civilian field, as a means for expansion and control of the market dynamics, or as a deterrent for the resolution of conflicts related to it, both locally and globally, on a small and large scale.

What is already being done is to prepare next generations of consumers willing, or predisposed, to become or to be transformed into one of the two types of cyborgs currently proposed by the high-tech communication marketing, the standard- integrated and the extra-integrated cyborg.

The standard-integrated cyborg (SIC) is the average technology products consumer affected by Technology Addiction or Communication Addiction, that thanks to a device inserted a dimora in his own body, a nanorobot or an intracranial microchip for neural interactions (brain-computer interfaces) for example [5] , or a subcutaneous chip (e.g. transponder implantation), can interface directly, via Near Field Communication (NFC) or via Far Field Communication (FFC) [6] , with an external device (e.g. a smartphone or a computer or any other wireless or bluetooth device equipped with an appropriate interface), or with the cloud (exposing himself to the risk of hacking).

This SIC will be easily identifiable and traceable and will be able to make some regular transactions, such as paying the bill at a supermarket or restaurant, make a withdrawal at an ATM, or punch the clock at entry and exit at workplace or school.

The extra-integrated cyborg (EIC), instead, is a man-machine hybrid trained to face high-risk experiences (e.g. military, policing or espionage missions, terrorist or criminal actions, space travels and stays in extra-terrestrial environments), supplemented by internal technological components which increase the self-adjusting body-mind functions (such as tolerance to pain, hunger and thirst, heat regulation, etc.), enhance the so-called normal performances (such as the ability to calculate, memorization and perceptual skills), make him or her remotely controllable and programmable [7] [8] [9] [10] [11] .

This is all that can be done and that is already being done, towards some next generations of technology products consumers affected by AI Addiction, a future littered with AI applications aimed at enhancing or inhibit (drugs-like?) certain functions or skills, in order to make the subject better performing and more competitive, or simply more monitorable.

APPROACHING INTELLIGENCE

The pendulum of the mind oscillates between sense and nonsense, not between right and wrong. C.G. Jung

Intelligence is a psychic, a mental or a cognitive function? In literature these three adjectives, psychic-mental-cognitive, are often used as synonyms, but in agreement with the interpretative model of the mental phenomenon which I have set in [1] [12] - [17] , it is appropriate to give them a distinct meaning. The reasons justifying the opportunity to introduce this distinction are three:

1) The first stems from the need to place the functions, in this case intelligence, in a phylogenetic perspective[6], trying to identify what can be the pre-constitutive psychological requirements, essential, predisposed ante rem, innate, phylogenetically prescribed, ready to use, that underpin the capacity, the possibility and the relational trend of the genus Homo, and on what structural and functional transformations and solutions, they went to meet during the long process of information of the inner life (insight) of human communities.

2) The second stems from the need to determine how these transformations and solutions have affected, over a period of time ranging from the Lower Paleolithic to the Upper Paleolithic, on the information of the functions.

3) The third stems from the need to clarify how and to what extent these functional transformations and solutions are correlated to the structural transformations and solutions that has faced the neuro-anatomy and/or the neuro-biology and/or the neuro- physiology of the CNS. In particular, we would like to clarify whether and to what extent there is a clearly established correspondence between the increase in functional capacities (e.g. Intelligence) and the increase of the cerebral mass in relation to body mass and/or the increase of the density of neurons and/or synaptic junctions in relation to the cerebral mass.

These three reasons lay down two distinct perspectives of investigation of the function. The first aims to explore the dynamics of the inner life of the genus Homo, that is its psycho-dynamics (the constitutive patterns, the trans-generational and supra-indi- vidual processes and transformations to which its insight went to meet over the phyletic time[7]). The second one focuses on the relationship of dependency between the psycho- dynamics and the structural and functional properties of the NS in general and of the CNS in particular, i.e. of the human neuro-dynamics.

The three acceptations, psychic-mental-cognitive, of the function intelligence will be discussed separately, first from the point of view of neuro-dynamics (Section 4 and 5) and then from the psycho-dynamics point of view (Section 6).

For convenience, the three definitions of the function referred to in section 6, will be provided at the end of that section. Advance use of the three adjectives meets these definitions, to which I refer.

Following (Section 7) will be provided a definition of intelligence that, in the absence of an established significant neuro-anatomical correlation, will be formulated within the framework of the interpretative physical model of the mental phenomenon which I have set in [Messori 2012B, 2013, 2015, 2016], that I'm going to summarize in Section 3.

As a general preamble to what I am going to expose, I would point out that from the perspective of this study, the phylogenetic changes, transformations and differentia- tions, genotypical and phenotypical, structural and functional, are never to be understood as the product of a purposeful design or plan, neither natural nor otherwise intended. On the contrary: 1) they are always the expression of solutions that thanks to an uninterrupted relationship of continuity, in time and space, are intertwined with all the solutions already adopted and with those adoptable (anticipatory systems), even when they are the result of a phylogenetic bifurcation[8]; 2) are always nonlinear and transient solutions (in relation to their evolution over phyletic times); 3) are always stationary but unstable solutions (in relation to the mutability of the unconditioned and conditioned habitat variables); are always intermediate solutions between adaptation and exaptation[9]; 4) are always solutions that for their becoming rely on the poietic action (which produces development and structure) exerted by the self-organization on the genesis and transformation of the forms of manifestation.

Continuity, transience, instability and non-linearity are properties that do not diminish but increase the possibilities of adaptation of biological systems to the continuity, transience, instability and non-linearity of the geo-environmental conditions, on a small and large-scale. Without them, for example, ecosystems would not have had, and would not have, any chance to re-invent and re-adapt themselves to the geo-environmental scenarios produced, e.g., by glaciations and by the various mass extinctions, which have followed in phyletic times as a consequence of natural disasters.

A scenario, the one that emerges from the perspective of investigation adopted here, in contrasts with the supposed quasi-finalistic and quasi-linear consequentiality of the phyletic lineage suggested by the Darwinian and neo-Darwinian arboreal model of phylogeny.

THE PHYSICAL DIMENSIONS OF THE UNIVERSE

Each element of reality, at any level of observation, is a micro system of relations, set in a macro system of relations. Nothing of what we observe exists in and by itself. Everything we can think of, observable and unobservable (i.e. virtual), instead of being made up of building blocks, or even quantized, in its ground state consists of tenso-relationship systems.

To clarify the meaning of these statements, and the effects they may have on the interpretation of the issues also examined in this work, we first need to set the ontological scenario within which to place the energy phenomenon, as well as the biological, neurological and mental one.

According to my Endo-Dynamo-Tensive Model [Messori 2012B], the establishment of a physical dimension based on a relational dynamic it arises as a consequence of a cosmogonic event, namely the supraliminal auto-perturbation (tensorial symmetry breaking) of a primary, irreducible and intrinsically dynamic state of tension[10], ontologically assumed as the physical background from which emerge all the differences in potential, all the interactions or forces and all the physical and psychical relationships. A second supraliminal perturbation (tensorial transition[11]) leads to the constitution of the spacetime system of correlative interactions between gradients of tension that we know as relativistic dimension or dimension of an exited tension- gradients distribution or ET-GD (mass and energy free dimension) (Figure 1).

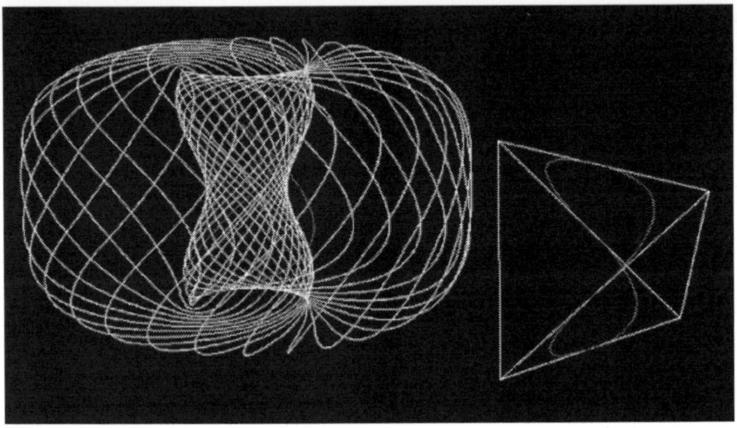

Figure 1. Image source: http://harmoniouspalette.com/TetMold.html. Biaxial or tetra-toroid, also coined as external toroid warped around an internal poloid, as drawn here has 27 identical loops. Compared with ordinary toroid coil, the main differences are twisted loops instead of the plain toroid loops and the involuted "donut hole". While 27 closed loops are presented to show a tetrahedron relationship, all loops can be one continuous twisting line. ET-GD structure corresponds to a non-local (diffuse) holographic cavity resonator with toroidal-poloidal topology consisting of a differentiated (tangent tensions → toroid) and undifferentiated (qualia → poloid) fractal- tensor foam, wrapped around a wormhole (vortex) lacking in structure and dynamics (emptiness/void/ vacuum). ET-GD structure constitutes the basic space-time fabric upon which are grafted all the physical and para-physical phenomena (psychism) that qualify the dimensions of the manifestation. All successive bifurcations are derived from this one via homomorphism (basic dynamic- structural equivalence) and translate the dynamic and structure of the ET-GD, by contextualizing it [Messori 2012B].

A third supraliminal perturbation leads to the constitution of a non-exited energy-quanta gradients distribution (zero point quantum field), from whose perturbation (supra-liminal oscillatory motions and/or charge densities of energy-quanta and impulse) takes shape the quantum and supra-quantum dimension (the dimension of the exited energy quanta-gradients distribution or EEQ-GD), where the relation (ship) is somehow explicated and may possibly be observed and directly experienced.

Every object or phenomenon of the quantum dimension is assimilable to a vibrational system (describable via a mathematical tool known as wave function), that vibrates with a certain frequential configuration, a certain oscillatory or phase modality (rhythm of oscillation) and a certain intensity, maintaining an uninterrupted local and non-local relationship of interference with other vibrational systems.

The phenomena of interference between the oscillatory modalities of the energy flows and impulse, involved in the perturbation/excitation of the quantum field, give rise to coupling-phase (oscillatory resonance[12]), able to trigger the phase transitions that lead, according to QED (Quantum Electrodynamic Field Theory), to the struc- turing of matter (Coherent Domains vs. Incoherent Domains). In particular, each localized (in space and/or in time) form of confinement (tensorial, energetic, massive, subatomic, atomic, supra-atomic, biological, cosmological), i.e. delimited by a boundary[13], is a tenso-vibrational micro-environment and corresponds to an oscillator or a resonant cavity (cavity resonator), a stationary system organized around a particular tensorial/frequential configuration of perturbations (tensions/oscillations), existing thanks to the relationships of interference it has with the endogenous and exogenous tenso-vibrational environment[14].

The final result is the diversification of the structuring of the phenomena affecting the exited energy quanta-gradients distribution (EEQ-GD) regime in four orders of phenomena, relatively autonomous and independent, associated with just as many physical varieties, the first belonging to the territory of the ET-GD, the second to the territory of the EEQ-GD, and the other two to the territory of the Hyper (cosmological) and Middle Dimension (H-MD):

- Tensorial phenomena (tensorial varieties: differentiated vs. undifferentiated tensions; qualia; images);
- Energy phenomena (electrodynamic varieties: wave fields and matter fields; wave- particle duality; anti-symmetrical/ chiral composite quantum states and symmetrical/achiral composite quantum states);
- Condensed matter phenomena (thermodynamic varieties, gas/liquid/solid, and chemical varieties, inorganic/organic);
- Biological phenomena (autopoietic varieties).

The condensative varieties of the EEQ-GD and of the H-MD derives from different correlations among tensorial gradients on the one hand, and, on the other hand, from different correlations among anti-symmetrical/

chiral composite quantum states (fermions[15]) and symmetrical/achiral composite quantum states (bosons), corresponding to different coherent oscillatory configurations (domains of oscillatory coherence or Coherent Domains, CD) that oscillate with a non-linear pattern to the rhythm impressed by a carrier frequency modulation.

On quantum and supra-quantum level, the weft of each energy phenomenon, condensed or rarefied, originate from the ongoing and self-organized warp of tensorial gradients constantly regenerated similar to themselves (in order to keep a global quasi- symmetrical condition) by the interaction between the quasi-continuous (fractal) and non-uniform distribution of monopolar/achiral gradients of non-hertzian potential (bosons → symmetrical wave function[16]; scalar component of the electromagnetic wave) and the quantized and non-uniform distribution of dipolar/chiral gradients of hertzian potential (fermions → anti-symmetrical wave function; vectorial component of the electromagnetic wave).

Furthermore, the four orders of phenomena relatively autonomous and independent affecting the EEQ-GD regime, can interact thanks to three types of correlative dynamics or couplings:

- Phase Conjugate Dynamics (of the type Frequency-Phase Correlative Dynamics);
- Spin Coniugate Dynamics (of the type Phase-Tension Correlative Dynamics);
- Tension Coniugate Dynamics (of the type Tension-Tension Correlative Dynamics)

However, the global balanced relationship that sets up this interfacing, namely the fact that the forms taken by energy-matter always remain similar to themselves (self similarity) is endowed by a cascade of unbalanced relationships. That is to say that locally (in space and/or in time) the universe of (self-similar) transformations, where nothing is created and nothing is destroyed, it is constantly shifted towards the component polarized in space or towards the component polarized in time. When this has to happen we speak of transitory quantum or supra-quantum symmetry breaking. Moreover, when this unbalanced state becomes stationary we may observe some very strange phenomenon, e.g. cosmological black holes when it is dramatically shifted towards the figures of interference polarized in space (space collapses), and living systems when it is shifted towards the quasi-tensorial figures of interference polarized in time. That is: in biological systems it prevails the component polarized in time and are themselves polarized in time.

According to this model, then, there aren't pieces of a mosaic (building blocks), but relationship systems (that in the ground state corresponds to pure tenso-relational systems). Their identity and their existence is determined by being part of a particular system of relationships, which manifest themselves solely in the correlative and functional availability of a complex series of energetic and/or tensorial relationships. Talk about hidden codes or pieces of thoughts (e.g., we have neurones that can generate electrical impulses, and these electrical impulses correspond to thoughts) or consciousness, without specifying that it is a scientific artifice, which can be provisionally adopted to facilitate the investigation of a certain object of study, it is a falsification of the mental phenomenon.

FROM PROTOPLASM TO NERVOUS SYSTEM: PART ONE

Living systems are transient systems tuned on state variations or stimuli (perturbative events as frequency variations, phase variations, tension variations) of the internal and external environment: their subsistence depends on their ability to adapt to (specific) applied perturbations. To adapt they must be, at first, selectively excitable. Selectivity is an essential condition for the existence of an autopoietic system. An example of biological selectivity is enantioselectivity. Biological systems are enantioselective, i.e. they strictly select the enantiomeric forms of the molecular species of which they consist (biological reactions synthesize and use always and only one of the two enantiomeric forms of a given molecule).

The enantioselectivity of biological systems is the reason for their homochirality, namely the presence of groups of molecules that have all the same enantiomeric configuration (e.g., the amino acids are all in the levorotatory configuration, while the ribose and deoxyribose of the nucleic acids have only dextrorotatory configuration).

Enantioselectivity, homochirality and autopoiesis of biological phenomenon seem to be related to the phase transition of water from the ordinary coherence of its liquid state to the semi-crystalline or glassy and super-coherent state of biological water [20] - [28] , the liquid medium within which the protoplasm (the colourless, translucent, viscoid, gelatinous and semi-fluid matrix of all cells) is formed.

The water (H_2O) is the third most common molecule in the Universe (following the H2 and CO molecule), and its chemical structure, based on the hydrogen bond, it is studied long since. Many models have been proposed but none has yet been able to explain a number of water properties considered anomalous [29] .

The Quantum Field Theory (QFT) and QED (Quantum Electrodynamic Field Theory) have produced a vision of water in a liquid state as a medium, which for a peculiarity of its molecular electronic spectrum reveals itself as an essential tool for long range communications, being able to change its supramolecular organization in function of the interaction with the environment.

The electromagnetic fields that are trapped within the Coherent Domains (CDs) of water and within their coherent matrices [30] , produce electromagnetic potentials that regulate the phase of the entire system, which in turn gives rise to selective attractions between the molecules of the solute.

"......the CD of liquid water", writes Emilio Del Giudice[17] [translation from Italian is mine], "unlike the CDs of other molecular species, is susceptible to give rise to a large number of excited states. Arises, consequently, the possibility of a further level of coherence, generated by the collective oscillation of a plurality of CDs of water, between two own configurations: a coherence between CDs, namely Super-coherence, that on one hand makes it grow the size of the related region by the tenth of a micron of the elementary CDs of water, up to microns of the cells, to centimeters of the organs, or to the meters of higher organisms. How it happens the collective oscillation of these CDs? Given the plurality of the excited levels of the CD, it is able to withdraw from the environmental noise small amounts of energy, transforming them into coherent vortices of quasi-free electrons. The duration of these vortical excitations can be very long (days, weeks, months), since because of the coherence the internal friction is zero, there are no collisions and the CD cannot dissipate energy in thermal form. Given the long duration of these excitations, it is possible to accumulate a large number of them within the domain, such as being able to accumulate a capital of 1 million euro in coins from 1 cent. Each vortex is a motion of electrons [31] [32] , that is, electrically charged particles, which gives rise to the appearance of a magnetic moment, which in turn aligns with the environmental magnetic field, which in the final analysis can also be the Earth's magnetic field. The vortices therefore cannot cancel each one another, but add up coherently, then transforming low-quality environmental energy (high entropy), in high quality coherent energy (low entropy), capable of inducing, as predicted by Szent -Gyorgyi, the electronic excitation of the molecules surrounding the CD. When the energy accumulated in the CD reaches the level corresponding to the energy of chemical activation of biomolecules present in the CD in the water, it develops, for electromagnetic resonance, the energy transfer [33] [34] [35] . The CD will discharge, thus completing its oscillation, and a set of well-defined chemical reactions take place, ruled by an electromagnetic resonance law, and not by the diffusive regime based on the random motions and collisions of molecules typical of non-biological chemistry. This makes it possible to achieve two objectives simultaneously:

- it allows the oscillation of the water CDs, a necessary precondition for the onset of the Super-coherence, extending the correlation radius to biologically relevant distances;
- it creates a biochemistry not governed by random collisions between molecules, but by a code of mutual recognition and recall among molecules based on long-distance electromagnetic interaction.

The biochemical reactions become therefore selective, that is obedient to specific organic codes, and fast. Furthermore, the chemical energy released during chemical reactions once taken from the CD modifies its characteristic frequency of oscillation, causing in turn the change of the type of molecular species able to resonate with the domain. We have, therefore, a non-stationary coherent regime, but that evolves over time, generating in turn a time-dependent biochemical framework. This way, we get a qualitative agreement with the main manifestations of the living dynamic, which appear incomprehensible in the conceptual framework of a chemistry ruled by diffusive regime and by the accidental collisions. The overall dynamics of the living appears to emerge, ultimately, by the oscillatory rhythm of water and by the related energetic events".

All biological envelopes, from cell membrane to epithelial tissue, contain this aqueous phase in a semi-crystalline state (likened to the liquid ice) or are perfused by it. It is water in a particular phase of quantum organization (oscillatory super-coherence) that gives it a high capacity to:

1) Retain electronic charges, in the form of vortical excitations of quasi-free electrons, storable as energetic reserve;
2) Induce an electronic and protonic long-range and long life excitation of the different molecular species available, enabling their selective activation and mutual attraction (→enantioselective absorption);
3) convert mechanical vibrations (phonons) in quantum of electromagnetic energy (photons) and vice versa (piezoelectric effect).

According to this line of research, the resonance phenomena induced by the electromagnetic field occupy a central role in the explanation of the establishment of the super-coherent phase of biological water, which in living organisms results also in plasma membranes, amniotic fluid (AF)[18], cerebrospinal fluid (CSF), interfacial and intracellular water (the viscous, glassy layers of interfacial water contiguous to biological surfaces are called Exclusion Zones, EZs, in which the solutes are not able to penetrate) (Figure 2), the latter being considered a key characteristic that has been exploited by Nature in setting up a mechanism able to match the quite different time scales of protein and solvent dynamics [36] , and a key characteristic for intercellular signal transmission [37] [38] [39] [40] .

From Protoplasm to Nervous System: Part Two

One of the general properties of the protoplasm, the biological material of which all cells are constituted (chemically composed by water, electrolytes, proteins, lipids and carbohydrates), possibly originated by the structuring process to which has undergone biological water [34] [41] [42] [43] during the early Archean period of the Precambrian Era, is irritability, namely the ability to sense an external stimulus and react to it. An organism consisting of a single cell (like the amoeba) reacts to a contact by contracting its cytoplasm. Irritability, however, is not an answer, but a cell must be irritable to respond to a stimulus.

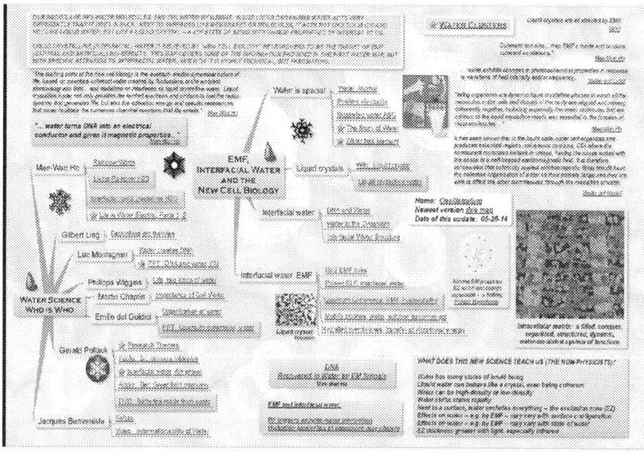

Figure 2. EMF, Interfacial water and the new cell biology. Image source: www.oscillatorium.com/id54.html.

In vertebrates, for example, epithelial cells, when stimulated, can respond by secreting, muscle cells by contracting, and nerve cells (neurons) producing and/or conducting neuro-electro-chemical impulses, or action potentials[19], that propagate along their extensions (axons and dendrites) at variable speed. In multicellular organisms of the animal kingdom (from Metazoa down to us), in which the various functions are distributed among different and increasingly specialized tissues, this characteristic property of irritability is acquired and perfected to the maximum degree by neurons that, together with other cells, such as glial cells [44] , form the Nervous Tissue (NT), whose morphological, anatomical and functional diversification and integration leads to the establishment of the Nervous System (NS) [45] [46] .

In animals with a structural and functional neurological degree of integration perfectly efficient and relatively diversified, such as Coelenterates, for example jellyfish and polyps, the NS is limited to a network of neuromuscular elements, equipped at the same time with sensitive and contractile capacity. In animals with a higher structural and functional degree of diversification and integration, as in some Shellfish, the NS is organized into more complex and selective structures, constituted by differentiated cells (that is, which have acquired particular structural characteristics fulfilling the function to be performed) grouped in depth, under the integuments, to form a chain of ganglia (neuronal clusters) at the side of the digestive tract. With the appearance of vertebrates the subdivision of the NS in central organs (notochord → neural tube) and peripheral structures (nerves and receptor organs), prefigured in some species of marine animals at the border between invertebrates and vertebrates (such as Pikaia gracilens, considered the fossil ancestor of the Amphioxus), undergoes a decisive stabilization and the central units of it (encephalus and spinal cord) are contained within a osteo-fibrous case represented by the skull and the vertebral canal.

Characteristic of the animals with bilateral symmetry (such as the species Homo) is a body structural plan arranged along a longitudinal axis, with the right half which is roughly the mirror image (but not overlapping) of the left half. An animal with bilateral symmetry can move more effectively than an animal with radial symmetry (such as Coelenterates) that, in turn, is better adapted to a sedentary life. Another characteristic of animals with bilateral symmetry is having a spatial orientation of the axial type, i.e. arranged longitudinally between two ends, a front (cephalic or rostral) and a posterior (caudal). Having a forward end is a characteristic of very active animals. In such organisms many of the sensory cells (nervous cells specialized in receptorial-stimulus specific functions) are concentrated in the front end of the body with a greater concentration of nervous cells specialized in functions other than those stimulus specific: thanks to this particular neuro-anatomical organization the organism is found to be practically oriented in a preferred direction, forward, on the basis of which the animal organizes its motor behavior.

Nervous System and Functional Role of Neurons and Glial Cells

The amazing thing about the brain is that it permits, for example, the sensory-motor co-ordination of the whole interaction, the hormonal regulation which ensures the maintenance of corporeal integrity, and so on, but the notion of neuronal correlates of mind as such is, to use the words of Alfred Norton Whitehead, "an inopportune concretization". Francisco Varela

In the non-linear process of phylogenetic diversification, nerve cells play the role of receptor units of zoological individual of tissue organisation, whose dual selective function on state variations (stimuli)

and functional interface between innervated tissues, supports and integrates the function exerted by the catalytic cellular core (CCC), formed by the Golgi apparatus, the centrosome (MTOC, Microtubule Organizing Center [47]) and microtubules (the structural units of the cell cytoskeleton, polymerized protein highly polarized), on energy-transfer. Cell differentiation which leads to the identification of the specialized cell neuron, responds to a phylogenetic request (→bi- furcation), which at a certain stage of evolution (Cambrian), gave a new degree of freedom to the biological systems in their relationship with the environment: the zoo-logi- cal line of neuro-dependent behavioural relationship.

Diversification ofbehavioral strategies require then cell differentiation, thc proccss by which cells specialize, acquiring or enhancing their ability to perform a specific function[20]. With cell differentiation the whole clusters of unicellular organisms that have colonized the planet until then, distributed over one or more areas and linked by a common adaptive and bio-energetic gain, found the way to come together (a process of in-formation governed by the self-organization of specific associative patterns induced, in the short and long range, by resonance phenomena involving various kind of cellular and molecular species) to become localized cells colonies, specialized and joined by a common structural and functional link, defined and identified in the construction of different tissues, organs and systems of organs, that operate coherently and in synchrony for the survival and unity of the multicellular system. The biological unit (cell), from being a composed, integrated system delimiting (membrane) a multitude of sub- cellular structures and molecular units, became an integrated specialized cellular ecosystem (multi-cellular organism), consisting of differentiated cells, specialized according to the role and function they must play to be part of a choral unit.

In vertebrates, CNS functional and morphological organization follows exactly this pattern. At increasingly specialized and diversified behavioral strategies, pointing at species-specific CNS functions, also corresponds a distribution of different, morphologically and/or functionally specialized neuron-glial colonies, whole groups of cells belonging to species-specific Coherent Domains of the CNS basin of attraction[21], each of them in a relation of continuity with all the others, where the single behaviour is prescribed by the colony's choral interference network, and described by a unanimous and synchronised dynamic, based on self-organized biophysical processes occurring via Phase Conjugate Dynamics, Spin Coniugate Dynamics, Tension Coniugate Dynamics.

Throughout the NT of invertebrates and vertebrates, neurons are invariably found in close association with glial cells (or glia-like cells in primitive animals, with very simple cellular organization), a type of cells that have been co-opted (exaptation), at least in vertebrates, to perform functions other than the original ones. Indeed, until recent times it was believed that only neurons and synapses were to carry out the functional activity of the NS and CNS, but increasing evidence accumulated in the last decades suggest that this role should be shared with glial cells, the number of which is significantly higher than that of neurons (ratio of about 10-50:1), which regulate the neuronal synaptic responses associated with learning and memory processes and share with neurons the role of mediators in the genesis of brain functional skills [48] [49] and occupy a prominent place in the neurogenesis of NT and in the evolution (e.g. via astroglyal lineage cells [50]) and architecture of the CNS[22]. The term glia includes non- neuronal cells known to play several functions in the maintenance of neural homeo- stasis in both central (CNS) and peripheral (PNS) nervous system. The main glial cell types are, in the CNS, astrocytes, oligodendrocytes and microglia and, in the PNS, Schwann cells, enteric glial cells and satellite cells. Evolutionary origins of glia are possibly linked to organizational needs, including structural support, by-product removal, phagocytic functions, developmental programming and circuit modulation [51] . Glia probably evolved from neuro-ectodermal cells, possibly from a mobile lineage or independently in different regions of the body, having multiple functions in both developing and mature nervous system [52] [53] . A notable exception is microglia, which are part of the immune system and enter the CNS from blood circulation early in an organisms development. Radial glia is generated before neurogenesis (see below), the process of generating functional neurons from neural and glial precursors (→ stem cells), very active in the pre-natal period (→ development of the CNS) and relatively active also in post-natal period (→ neuronal repair processes) [54] , and guide neuronal migration in the developing brain. In mammals, radial glial cells disappear or become astrocytes within days to weeks after birth, but in non-mammals they persist into adult- hood as radial glia [55] [56] .

Neurogenesis is a complicated process of generating functional neurons, which involves, in addition to neural and glial precursors, mature glial cells, the CSF and epen- dymal cells[23]. During embryonic development, the vast majority of neurons and a large number of glial cells are generated in the germinal zones or germinal matrices or Neurogenic Niche [57] [58] [59] , the micro-anatomical environment that depending on the period of neuronal development (prenatal → postnatal) may contain endothelial cells, ependymal cells, stem cells, astrocytes, microglia, neurons and mature descendants of adult neural precursors, and that functionally controls their development in vivo. In the early phases of cytogenesis, the germinal zones are located primarily at the ventricles surface (ventricular zone, VZ). While embryonic development proceeds, on the surface of VZ forms a germinal subventricular zone (SVZ)[24], which together form the ventricu-lar-subventricular zones (V-SVZs) [60] [61] [62] .

As the cerebral hemispheres enlarge, and the distance to be travelled by the cells increases, most of the neurons that will form the cerebral cortex migrate to their destinations along specialized radial glial fibers (RGFs) that span the entire thickness of the hemisphere from the ventricular surface to the pia[25].

After birth neurogenesis is carried out by the specialized activity of the neurogermi- native subgranular zone (SGZ) of the dentate gyrus of the hippocampus [59] [63] [64] .

THE AUTOPOIETIC PSYCHO-NEURO DYNAMICS

I view the brain not as a box with compartments that contain sadness, joy, color, texture, and all the other "objects" and categories that one might think of. Instead, I envisage it as a constantly shifting dynamic system; more like the flow of a river in which patterns emerge and disappear, than a static landscape. This is an entirely different image from the brain as a computer with stored contents or subroutines to be called up by a program. In nature's pattern-forming systems, contents aren't contained anywhere but are revealed only by the dynamics. Form and content are thus inextricably connected and can't ever be separated. J.A. Scott Kelso

Through NT and NS, the organism may develop the ability to interface biunivocally the territory of tension with the territory of energy, so as to give way to new relational and adaptive strategies (new compared to the non- or pre-neurological relational and adaptive strategies). In this sense, NT-NS as a unit is a selective interface between the two territories, energy ↔ tensorial, and the original function assigned by cell differen- tiation to nerve cells is receptorial and meta-physical (of interfacing between one plan and the other of reality): that is, NT and the NS, regardless of the number of anato- mical/functional units (neurons, glial and the like) involved, are condensed (fermionic) systems specialized in interfacing the Exited Tension-Gradients Distribution (ET-GD) with the Exited Energy Quanta-Gradients Distribution (EEQ-GD), and viceversa, for adaptive purposes. Photons (massless objects) and phonons (quantized mode of vibration) exchanges generated by the neuro-electro-chemical activity are at the basis of this dynamic of interfacing.

With respect to the territory of energy, NT-NS makes a selection (receptor function) species-specific upon the state variations or stimuli (perturbative events, e.g. frequency variations, phase variations, tension variations) from the external and internal environment, making the organism appropriately sensitive to a certain range of state variations and not to others. Moreover, NT-NS is specialized in transforming, upon kinematic similarity, the chemical and physical state variations of the environment, in spatial and temporal neuro-electro-magnetic phase correlations. With respect to the territory of tension NT-NS it transmutes the stimuli from frequencial pattern (receptor activity↔neu- ro- electro-chemical activity) in isomorphic pattern of spin, on which operates a specific Holographic Strange Attractor[26] [12] [13] [14] , which transmutes the pattern of spin in tensorial pattern, and viceversa, and sets the functional correlations between them, valid under the biological and/ or neurological and/or psychic standpoint.

The mind(body) system, i.e. the psycho-neuro dynamics characterizing an alive and vibrating human being (all neurologic organisms exhibit proto-mindedness, but not all of them are minded [65]), it can be seen as an autopoietic system strung between the Exited Tension-Gradients Distribution (ET-GD) and the Exited Energy Quanta-Gradients Distribution (EEQ-GD), governed by the syntropic action of an Holographic Strange Attractor, able to develop different degrees of integration and to carry out different functions, that responds to the laws of psychodynamics of non-equilibrium, that constantly redefines itself and that internally it sustains and reproduces itself. The autopoietic dynamic of mind (body) system is organized around biophysical autocatalytic patterns (i.e. self-accelerated) regulated by continuous and non-linear fluctuations of selective tension transfer between intra and extra environment. The trans-neurological selectivity is the central element of the autopoietic psycho-neuro dynamics. There is a diffuse and integrate neurological intercellular relationship, supported by a dense network of physical and chemical signals (via synapses, gap junctions, intra- extracellular fluid, plasma membranes, CSF [66] [67]) capable of interacting with the environmental substrate so as to produce the net of soliton waves, i.e. neuro-electro-chemical impulses or nerve impulses or action potentials, that qualifies the sensory/receptor system activity. Selective tension transfer between intra and extra environment is associated with the production and transduction of tensorial signals (coherent scans of tension state variations or tenso- stimuli) that come into play in the tensorial interconnection and in the biophysical processes of Nervous System tuning (via Phase Conjugate Dynamics; Spin Coniugate Dynamics; Tension Coniugate Dynamics): that is, all psyco-neuro information dynamic is carried, transformed and made available as holographic spin-wave interference patterns on the surfaces of higher order hyperspace fields (e.g. on a hypersphere, a structure embedded in a 4D space, equipped with a double donut-like shape, which is invisible in the usual 3D dimensions [68] [69] [70] [71]), which are non-locally resonant with the intermediate EM field generated by the CNS activity (Figure 3).

In general, a Nervous System defines and separates this network of interactions from the environment so that it can realize an autonomous but not isolated unit, namely a living neurologic organism. Humans, and to some extent also other vertebrates, are engaged in engramming a sensorineural warp of tensions on which it can weave the weft of a variably integrated state of mind.

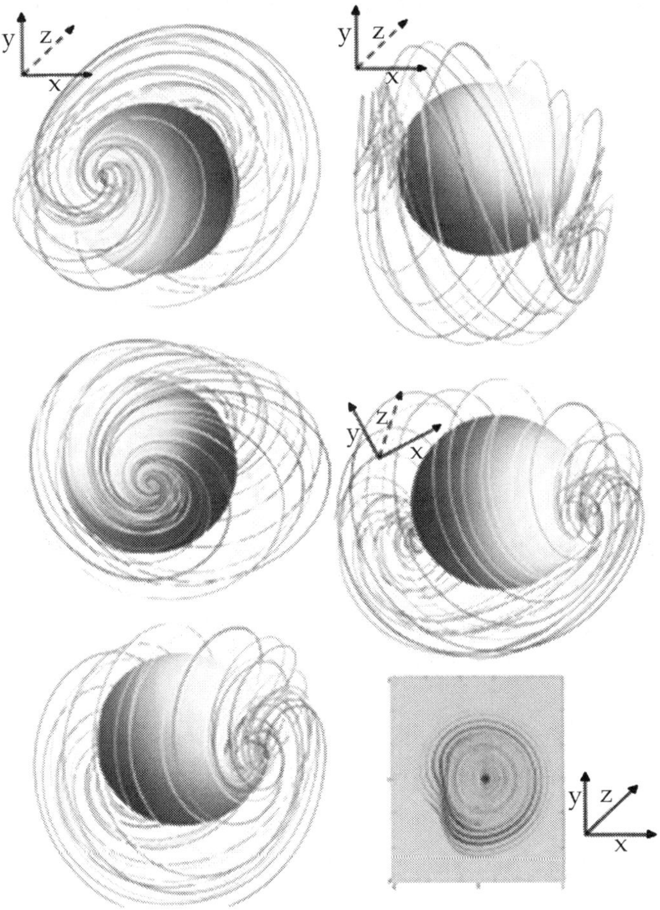

Figure 3. Image source: http://arturotozzi.webnode.it/products/strange-attractors-and-tori-embedded-in-a-brain-phase-space/. STRANGE ATTRACTORS AND TORI EMBEDDED IN A BRAIN PHASE SPACE. The brain is a system at the edge of chaos equipped with nonlinear dynamics and functional energetic landscapes.

In vertebrates, the sensorineural response common to all neurological organisms is processed and integrated, e.g. via super-complex anticipatory systems[27], to become perception (perception being a function more integrated than sensitivity which is a function more integrated than reacting which is a function more integrated than excitability), which in humans is associated to sensing, the mental substrate on which thinking is grafted (see next Section).

PSYCHO-DYNAMICS OF THE PSYCHIC-MENTAL-COGNITIVE FUNCTIONS

Like a river whose eddies, vortices, and turbulent structures do not exist independent of the flow itself, so it is with the brain. Mental things, symbols and the like, do not sit outside the brain as programmable entities, but are created by the never ceasing dynamical activity of the brain. The mistake made by many cognitive scientists is to view symbolic contents as static, timeless entities that are independent of their origins. Symbols, like the vortices of the river, may be stable structures or patterns that persist for a long time, but they are not timeless and unchanging. J.A. Scott Kelso

From the Lower Paleolithic to the first half of the Middle Paleolithic, the action produced by the encounter between human intelligence and the environment was limited to ensure the performance of adaptive phylogenetically prescribed functions, with few but significant deviations from the phylogenetic prescription, which already from the first half of the Lower Paleolithic attest to how the human species were phylogenetically endowed with the possibility to overcome the behavioral stereotypy, to which are irreducibly bound all other animals [Messori, 2016]. Throughout this period, about two million years, the psychic basin of attraction around which it has revolved the human psychodynamics, was essentially composed by what C.G. Jung calls sensing-intuition bipolar psychic dimension[28], an irrational psychic dimension composed by the psychic function sensing, that due to the action carried out by the six sense organs[29] allows the individual to maintain an adequate state of attention to, and behave depending on, the environmental stimuli (energy and tensorial state variations), and the psychic function intuition, which it provides the individual a pool of raw psychic contents, prescriptive but not descriptive (archetypes[30]) that lend themselves to nourish, and to be fertilized by, the relationship with the environment. With the psychological birth of human individuality (Middle Paleolithic) [Messori 2016] and the subsequent acquisition of a distinct identity from the environment (relationship of contiguity), indispensable prerequisite for the establishment of what we call self-consciousness[31], the psychic environment changes radically its structure and the bipolar psychic dimension sensing ↔ intuition is integrated with another fundamental bipolar psychic dimension, in this case rational, composed by the psychic→mental function thinking, that through recourse to logical processes tend to derive mind-facts from the raw object (archetype), and the psychic→mental function feeling, which tends to assign a value to things depending on the characterial predispositions and expectations of individuals who experience it[32].

From this time onwards the human being begins to entertain a dialogical relationship with the environment, mediated by the state of tension that develops between the phylogenetic content (→independent variable), which gives-form (in-form) to the bipolar psychic dimension sensing↔intuition and the epigenetic content (→dependent variable), which in-form the bipolar psychic dimension thinking↔feeling, and to identify itself as an agent of mediation between Earth and Sky, between the consistency of a reality perceived through the six sense organs, and supported by the psychic function sensing (→Earth), and the consistency of a reality intuited through archetypal psychic contents, and supported by the psychic function intuition (→Sky)[33].

Animated by the dynamics of the new psychic → mental environment, the exercise of intelligence comes out from the incubation stage where has operated for about two million years, entering a phase in which the epigenetic learning process (→dependent variable) integrates and shapes the phylogenetic prescriptive action (→independent variable), bringing to maturity the formation of mental functions and adaptive and supra-adaptive skills that can also make use of relational modules other than the one (phylogenetically prescribed) that has characterized the social life of human communities from Lower Paleolithic to Middle Paleolithic, i.e. different from the imitative module, based on the imaginific function[34] (→ bipolar psychic dimension sensing ↔ intuition) and mediated by the human acoustic-musical faculty[35] [Messori 2012A].

Between Middle and Upper Paleolithic, the action carried out by the imitative module in the relational life of human communities of Homo Sapiens et Faber was built, initially, by the symbolic module and then (from Upper Paleolithic onwards) by the ideative module, based on the namesake functions, now yes, cognitive.

Based on these premises, the psychodynamic distinction that can be made between psychic function, mental function, cognitive function, is the following:

- The psychic function is the relational capacity, distinctive, involved in the phylogenetically prescribed arrangement of the human psychic environment (insight), i.e. its species-specific tensorial basin (the particular tensorial warp on which it can be weaved the energy weft characterizing the genus Homo) (Figure 4) in whom are dissolved all the phase transitions (bifurcations) who went to meet the psychism[36] (C.G. Jung) of the biological phenomenon in the course of phylogeny [Messori 2000, 2012A, 2015, 2016], passing through the neurological transition up to the bifurcation that led to the emergence of the genus Homo.
- The mental function is the explicated, modal and procedural, form of the psychic function. It refers to the process of information that went to meet the human psychic → mental environment in the relationship with the environment.

De facto, the psychic function and the mental function have always been compresent (what varies is the prevalence of one over the other), and they point to the Jungian concept of archetype (psychic raw material) and to that of collective unconscious.

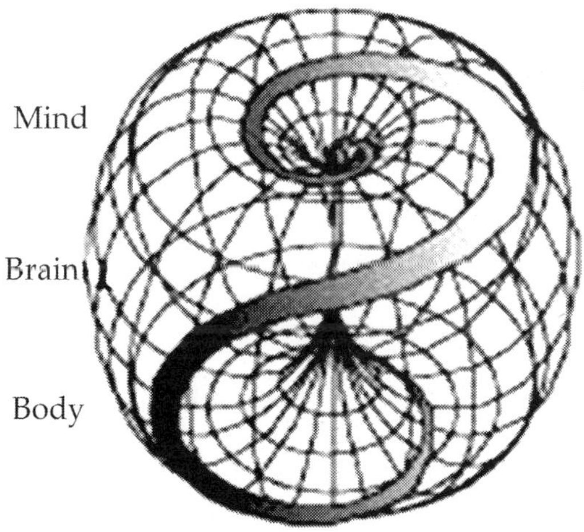

Figure 4. The human tensorial basin (of attraction), i.e. the mind-brain-body (de-localized) system, reproduces the fundamental structure characterizing the EEQ-GD, which in turn reproduces the fundamental structure of the ET-GD. The fundamental structure of the EEQ-GD is a unitary fractal and resonant object, a toroidal-like cavity resonator, consisting of an outer frequency weft

(Phase Conjugate Dynamics → Body), an inner spin weft (Spin Coniugate Dynamics→ Brain) and a tensorial nucleus (Mind→ Tension Coniugate Dynamics). On the ET- GD level, the precursor of this fractal object, is a diffuse toroidal vortex consisting of spin-inter- nal motion (zitterbewegung, self-dynamism not dependent upon external factors), a stationary and fractional (non-quantized) entity within which the motion impulse is only latent. Subjected to a torque and rotational action (induced by reflection symmetry breaking) this diffuse toroidal vortex transform into a bi-polar quantum-relativistic object, named Twisted-Pinched Hysteresis Loop (T-PHL) [Messori 2012B], in which the spin-internal motion is de-symmetrized in spin-up and spin-down. The final outcome (EEQ-GD) is a continuous and non-uniform cyclotronic motion of dextrorotatory vs. levorotatory spin-angular momentum, that in passing through the de-loca- lized point of self-intersection of the loop (the eight-like strip highlighted in the picture) decussate and reverses sign, moving seamlessly from negative values to positive values. The mirror symmetry breaking leading to the biological T-PHL is at the basis of the enantioselective absorption mediated by biological water, which determines biological homochirality.

From the psycho-dynamic point of view, these two functions qualify the constitutive arrangement, and the dynamics, of the psychic → mental environment, which precedes the psychological birth of the genus Homo (Middle Paleolithic).

-The cognitive function, is the relational capacity, gained from the particular differentiation, and subsequent arrangement, of the psychic → mental environment, after the psychological birth and, in particular, by the development of oral language and by the maturation of cognitive thinking (→ bipolar psychic dimension thinking-feeling → logical-abstract thinking), understood as the ability to program significant actions with respect to the ability to formulate meaningful purposes, by resorting to the exercise of ideation and ratiocinative discrimination.

INTELLIGENCE

In those places, the nicknames, the jungle names, were defined by psychologists as "construction of a double", the purpose of which was to dissociate children from guilt and totally detach them from their self prior to recruitment; in Sierra Leone some childish units called themselves "cyborgs" denoting consciously themselves as semi-human machines, devoid of feelings. Alberto Sanza

The phylogenetic history of the genus Homo, of which we can claim to know, in its fossil record, at least five species of which only one, Homo Sapiens, currently living, began about 2.5 million years ago with Homo Abilis. To date, the study of brain evolution in the fossils it has been restricted to the analysis of the variations occurred over the phyletic time 1) by the cranial capacity, 2) by the surface layers of the cerebral volumes, such as the marks left on the intracranial wall by the cerebral convolutions and by the vascular structures (arteries and veins) that have been shaping the surface of the cerebral cortex, and 3) by the bony structures that serve as docking or anatomic reference for the soft tissues, all variables that are suitable to be used in studies of comparative paleoncuro-anatomy, but that does not give any information on the deep areas of the brain, and especially do not allow to trace the phylogenetic evolution of the psychic functions. In recent times, the transition to digital morphology and multivariate analysis techniques, interpreted according to the principles of morphological integration, has increased dramatically the possibility of anatomical reconstruction, allowing the study of intracranial form (endocast[37]) to go well beyond the simple analysis of brain volume. Many encephalic variations can therefore be interpreted as non-neural consequences of encephalization process (increased cranial capacity in relation to body size), or as structural consequences of changes associated with the facial structures or with those at the base of the skull. Other changes are more directly interpretable as real changes in cortical areas, with possible functional relationships of cognitive character. Specifically, the morphological changes in the higher areas of the brain (frontal and parietal lobes) are typically more sensitive to real neural changes. Changes in the conformation of the frontal and parietal lobes are described in the most encephalized human species, i.e. in Neanderthal populations and in anatomically modern humans. In the latter, the changes of the parietal volumes are extremely noticeable, and are also associated with marked changes in cerebral vasculature. In return, the lower brain structures (temporal, occipital and cerebellar lobes) are subjected to biomechanical stresses of the base of the skull, which often determine changes in its shape, without necessarily affect neural functional changes. In any case, there are not elements to establish a deterministic and linear correspondence of cause and effect between the diversification of brain functions and the morphological and structural changes which went to meet the CNS in vertebrates. Moreover, there aren't elements to establish a relationship of proportionality between the encephalization index and the quality or quantity of brain functions expressed by an organism, just as there is no proportional relationship between the number of its genes and the quality or quantity of its behavioral strategies[38].

Furthermore, since the relationship between mind and brain, i.e. between psyco-dynamics and neurodynamics, it has nothing to do with transmission of information, exchange of meaningful messages, decryption of encrypted code, computational encoding of bit of information and the like (if the genesis of the biological and mental phenomena had to be based on the precepts of Information Theory, there would be no living nor minded systems at all), the definition of intelligence that follows is formulated according to the psycho-dynamics line of investigation.

The word intelligence comes from the Latin intellegentia (or intelligentia), which in turn comes from the verb intellĕgo (o intelligo), derived from lĕgo with the addition of the prefix, which is also a preposition, inter, "in the middle", "between" (or of the adverb intus, "inside"). The Latin verb lĕgo is related to the Greek verb légo with which it shares the same Indo-European root *LAG-/*LEG-, which it expresses the idea of harvest, collect, gather, pull together and tune different things, connect, from which the Sanskrit lagati, "to bind to", "to attach". From this root, the Greek légo has developed the lexical form of the "collecting-braiding-talk" (meaning "to collect, braiding the sounds and tune the words in a phrase"), that refers to the dichotomous nature of the lógos, the ordering function of the lógos as "logic-word" (speech, truth, principle, law, reasoning, discernment), and the function as medium of the lógos "magic-word" (intelligible sound, en-chant, mythos, mantra, symbol), while the Latin lĕgo has developed the lexical form of the "collecting-braiding-read" (meaning "collecting with the eyes the graphic signs of the words, pronouncing them within onself, in silence"), from which lex, "the law" (as code that collects and concretises the general principles of the ius), and re-ligio, "religion" (renew the bond with alterity).

From this etymological reconstruction it turns out that, in summary, intelligence indicates the ability to select "between and inside", weaving and tuning to each other and within them different things.

Keeping this definition and developing it so as to obtain a sufficiently articulated and unequivocal definition, we can say that: with the term intelligence we refer to the capacity, congenital and acquired, conscious and unconscious, of an individual or a community of individuals, to select, tune, polarize and apply his/her/their psychic → mental → cognitive functions, in the identification and/or in carrying out the behavioral and/or knowledge strategies, adaptive or supra-adaptive, utilitarian or non-utilitarian, as a response, genetic or epigenetic, to the relational and/or knowledge needs that arise by the encounter (made of dependent and independent, predictable and unpredictable variables) between the pressure exerted by the inner environment (insight) and the pressure exerted by the external environment.

As such, intelligence does not indicate the abilities for this or that specific relational skills, or competence or psychic-mental-cognitive function, neither it refers to the degree of development of one of them, nor can be broken down into a more or less extended series of capacities. It makes no sense, rather it is patently incorrect, to speak of logical-mathematical intelligence, verbal i., spatial i., musical i., kinesthesic i., emotional i., etc. Intelligence is a global, choral, systemic, property not sectorizable nor factorizable, which reveals itself in the encounter between the action taken by the individual/ community into the environment and on the environment, and the action exerted by the environment on the individual/community and in the individual/community, a global property that depends on the possibilities, capacities and tendencies of those who exercises it, in front of all the variables involved.

Is it possible to relate or even identify intelligence with the computing and storage capacities of an intelligent machine? What can it have in common with the artificial, computational, predictable, reversible, reproducible, planable and controllable skills' simulation carried on by the ASSS?

CONCLUSIONS

Western man has no need of more superiority over nature, whether outside or inside. He has both in almost devilish perfection. What he lacks is conscious recognition of his inferiority to nature around him and within him. He must learn that he may not do exactly as he wills. If he does not learn this, his own nature will destroy him. He does not know that his own soul is rebelling against him in a suicidal way C.G. Jung

The epigenetic learning makes use of behavioral strategies and tools of investigation and knowledge that since the Bronze Age are placed and proceed on two different but complementary logic levels, one rational and the other irrational [1] [72] . One is dynamic, facing outward, oriented to the elaboration and adoption of strategies and tools in knowledge of the physical, material, formal, sensible, immanent reality, circumscribed by the dimension of rationality and meanings. When intelligence and learning are developed on this plan, the skills that are acquired are supported predominantly by the thinking-feeling bipolar psychic dimension, and meet practical, utilitarian, govern- ment-like aims. The other is static, facing inward, oriented to the elaboration and adoption of strategies and tools in knowledge of the immaterial, informal, meta-physical, magical reality, circumscribed by the dimension of irrationality and symbols (the excess of meaning, the alterity). When intelligence and learning are developed on this plan, the skills that are acquired are supported predominantly by the sensing-intuition bipolar psychic dimension, and can lead to two different solutions: one makes use of strategies and exterior tools of investigation mediated by the senses, but it is devoid of practical and utilitarian purposes (e.g. the artistic production), while the other is totally detached from both practical and utilitarian purposes and from strategies and external tools of investigation mediated by the senses (ecstatic experiences, contemplative states, deep meditative states).

All civilizations born out of metallurgy revolution had to, have to and will have to deal with these two logic levels, from whose elaboration takes shape their respective internal representation of external reality, their culture, social and cultic organization.

That is not by chance that in the last years of his life, Auguste Comte (1798-1857), the ideologist of positivism, writes the positivist Catechism and founded the Positivist Church, where are transposed the doctrinal, ethical and liturgical elements of the Catholic Tradition. In fact, by doing so, he closes the circle around his vision, positivist, of the metaphysic applied for practical purposes (Messori 2015), the same vision that animates contemporary scientist and technicians, who want to emulate the theologians that almost two thousand years ago managed to bundle the alterity in the consubstan- tiality of the Christological perichoresis, i.e., they made immortal what is mortal and mortal what is immortal, an alchemical operation remained unequaled, at least until today, the

Age of the Super-Enlightenment, in which science seeks to give life and consciousness to the machines!

Who knows, maybe to realize how silly it is to describe biological minded systems adopting the criteria applied by techno-science to build and program the so-called intelligent artificial systems, i.e. idiot machines, claiming that we should learn from their intelligence; meaning that we have to be idiots as they are, and we'll have to wait for a child to be accompanied by parents at the opening party of the 2040 AI Olympic Games, and while the crowd cheers all around, the puzzled child asks to the mother: Mom, but do they know that those over there are just machines?

REFERENCES

1. Messori, C. (2016) From Continuity to Contiguity. On the Genesis of Consciousness, Culture and Oral Language—Part I of IV. Journal of Consciousness Exploration & Research, 7, 163-228.
2. https://www.researchgate.net/publication/295857703_From_Continuity_to_Contiguity_On_the_genesis_of_consciousness_culture_and_oral_language_Part_I_of_IV
3. Messori, C. (2000) Il Sole e la Luna. Sulla Natura dei Simboli e della Mente Umana. Federico Ceratti Editore, Milan.
4. Messori, C. (2012) Dalla Facoltà Acustico-Musicale alle Origini del Linguaggio Orale Fino al Predominio della Cavità Orale che Genera il Mondo sulla Cavità Uterina che Genera la Vita. Il Minotauro, Persiani Editore, Bologna, 2, 6-43.
5. Mencattini, A., et al. (2014) Speech Emotion Recognition Using Amplitude Modulation Parameters and a Combined Feature Selection Procedure. Knowledge Based Systems, 63, 68-81. http://dx.doi.org/10.1016/j.knosys.2014.03.019
6. https://www.researchgate.net/publication/261328157_Speech_emotion_recognition_using_amplitude_modulation_parameters_and_a_combined_feature_selection_procedure
7. Fleischer, J.G. and Edelman, G.M. (2009) Brain-Based Devices: An Embodied Approach to Linking Nervous System Structure and Function to Behaviour. IEEE Robotics & Automation Magazine,

16, 33-41. http://dx.doi.org/10.1109/MRA.2009.933621. http://www.nsi.edu/~fleischer/fleischer_edelman_ram.pdf

8. Agbinya, J.I. (2011) Principles of Inductive Near Field Communications for Internet of Things. River Publishers.
9. Regalado, A. (2014) Military Funds Brain-Computer Interfaces to Control Feelings. MIT Technology Review, May 29. https://www.technologyreview.com/s/527561/military-funds-brain-computer-interfaces-to-control-feelings/
10. Strickland, E. (2014) DARPA Project Starts Building Human Memory Prosthetics. IEEE Spectrum, August 27. http://spectrum.ieee.org/biomedical/bionics/darpa-project-starts-building-human-memory-prosthetics
11. DARPA (2015) Neurotechnology Provides Near-Natural Sense of Touch. DARPA News and Events. www.darpa.mil/news-events/2015-09-11
12. Sanchez, J. (2015) Restoring Active Memory (RAM). DARPA Program Information. www.darpa.mil//program/restoring-active-memory
13. Sanchez, J. (2015) Systems-Based Neurotechnoloy for Emerging Therapie (SUBNETS). DARPA Program Information.
14. Messori, C. (2011) Cells, Neurons, and Qualia: The Holographic Strange Attractor Model. Journal of Consciousness Exploration & Research, 2, 1417-1437. https://www.researchgate.net/publication/255696668_Cells_Neurons_and_Qualia_The_Holographic_Strange_Attractor_Model
15. Messori, C. (2012) A Cosmogonic Model of Human Consciousness. Journal of Consciousness Exploration & Research, 3, 1149-1208. https://www.researchgate.net/publication/255696726_A_Cosmogonic_Model_of_Human_Consciousness
16. Messori, C. (2013) L'Antropologia della Coscienza incontra la Fisica Quantistico-Relativistica: Modello Cosmogonico Endodynamotensivo (EDT) e Mnemopoiesi (MOPS). Volume 1, Il Minotauro, Persiani Editore, Bologna, 1, 7-82.
17. Messori, C. (2015) Quale Modello per le Neuroscienze—Part. I of III. Volume 1, Il Minotauro, Persiani Editore, Bologna, 1, 25-45.

18. Messori, C. (2015) Quale Modello per le Neuroscienze—Part II of III. Volume 2, Il Minotauro, Persiani Editore, Bologna, 2, 80-117.
19. Messori, C. (2016) Quale Modello per le Neuroscienze—Part III. Volume 2, Il Minotauro, Persiani Editore, Bologna.
20. La Marr, W.A. (1998) The Effect of Supercoiling on Small Molecule-DNA Interactions. Massachusetts Institute of Technology, 1-103. https://dspace.mit.edu/bitstream/handle/1721.1/50414/41163964-MIT.pdf?sequence=2
21. Vologodskii, A.V., et al. (1979) Fluctuations in Superhelical DNA. Nucleic Acids Research, 6, 967-982. http://dx.doi.org/10.1093/nar/6.3.967. http://www.ncbi.nlm.nih.gov/pmc/articles/PMC327745/pdf/nar00444-0159.pdf
22. Zhou, E.H., et al. (2009) Universal Behavior of the Osmotically Compressed Cell and Its Analogy to the Colloidal Glass Transition. Proceedings of the National Academy of Sciences of the United States of America, 106, 10632-10637. http://dx.doi.org/10.1073/pnas.0901462106
23. De Ninno, A., Del Giudice, E., Gamberale, L. and Congiu Castellano, A. (2014) The Structure of Liquid Water Emerging from the Vibrational Spectroscopy: Interpretation with QED Theory. Water Journal, 6, 13-25. http://www.waterjournal.org/uploads/vol6/deninno/WATER.2013.13.DeNinno.pdf
24. Ho, M.W. (2014) Large Supramolecular Water Caught on Camera: A Review. Water Journal, 6, 1-12. http://www.waterjournal.org/uploads/vol6/ho/WATER.2013.12.Ho.pdf
25. Clegg, J.S. (1982) Alternative Views on the Role of Water in Cell Function. In: Franks, F. and Mathias, S.F., Eds., Biophysics of Water, John Wiley and Sons, New York, 365-383.
26. Antonenko, Y.N., Pohl, P. and Rosenfeld, E. (1996) Visualisation of the Reaction Layer in the Immediate Membrane Vicinity. Archives of Biochemistry and Biophysics, 333, 225-232. https://www.jku.at/biophysics/content/e54633/e54639/e54665/
27. Mollenhauer, H.H. and Morré, D.J. (1978) Structural Compartmentation of the Cytosol: Zones of Exclusion, Zones of Adhesion, Cytoskeletal and Intercisternal Elements. In: Roodyn,

D.B., Ed., Subcellular Biochemistry, Vol. 5, Plenum Press, New York, 327-362.

28. Barry, P.H. and Diamond, J.M. (1984) Effects of Unstirred Layers on Membrane Phenomena. Physiological Reviews, 64, 763-872.
29. Pollack, G.H. and Clegg, J. (2008) Unexpected Linkage between Unstirred Layers, Exclusion Zones, and Water. In: Pollack, G.H. and Chin, W.C., Eds., Phase Transitions in Cell Biology, Springer Science & Business Media, Berlin, Germany, 143-152.
30. Tedeschi, A. (2010) Is the Living Dynamics Able to Change the Properties of Water? International Journal of Design & Nature Ecodynamics, 5, 60-67. www.witpress.com/Secure/ejournals//papers/D&NE050108f.pdf
31. Keutsch, F.N. and Saykally, R.J. (2001) Water Clusters: Untangling the Mysteries of the Liquid, One Molecule at a Time. Proceedings of the National Academy of Sciences of the United States of America, 98, 10533-10540. http://www.pnas.org/content/98/19/10533.full.pdf
32. Murugan, N.J., Karbowski, L.M. and Persinger, M.A. (2014) Serial pH Increments (~20 to 40 Milliseconds) in Water during Exposures to Weak, Physiologically Patterned Magnetic Fields: Implications for Consciousness. Water Journal, 6, 45-60. http://www.waterjournal.org/volume-6/persinger-summary-2
33. Meyl, K. (2012) DNA and Cell Resonance: Magnetic Waves Enable Cell Communication. DNA and Cell Biology, 31, 422-426. http://dx.doi.org/10.1089/dna.2011.1415.http://www.k-meyl.de/go/Primaerliteratur/Magnetic_Waves-Enable-Cell_Communication.pdf
34. Meyl, K. (2012) About Vortex Physics and Vortex Losses. Journal of Vortex Science and Technology, 1, Article ID: 235563. http://omicsonline.com/open-access/about-vortex-physics-and-vortex-losses-2090-8369.1000101.pdf?aid=15110
35. Del Giudice, E. and Tedeschi, A. (2009) Water and the Autocatalysis in Living Matter. Electromagnetic Biology and Medicine, 28, 46-54. http://dx.doi.org/10.1080/15368370802708728.http://www.ncbi.nlm.nih.gov/pubmed/19337894

36. Voeikov, V.L. and Del Giudice, E. (2009) Water Respiration—The Basis of the Living State. Water Journal, 1, 52-75.. www.waterjournal.org/uploads/vol1/voeikov/WATER-Vol1-Voeikov.pdf
37. Brizhik, L.S., Del Giudice, E., Tedeschi, A. and Voeikov, V.L. (2011) The Role of Water in the Information Exchange between the Components of an Ecosystem. Ecological Modelling, 222, 2869-2877. http://dx.doi.org/10.1016/j.ecolmodel.2011.05.017. https://www.researchgate.net/publication/229327908_The_role_of_water_in_the_information_exchange_between_the_components_of_an_ecosystem
38. Pagnotta, S. and Bruni, F. (2007) The Glassy State of Water: A "Stop and Go" Device for Biological Processes. In: Pollack, G.H., et al., Eds., Water and the Cell, Springer Verlag, Heidelberg, German, 93-112.
39. Agnati, L.F., et al. (1995) Intercellular Communication in the Brain: Wiring Versus Volume Transmission. Neuroscience, 69, 711-726. http://dx.doi.org/10.1016/0306-4522(95)00308-6
40. Agnati, L.F. and Fuxe, K. (2014) Extracellular-Vesicle Type of Volume Transmission and Tunnelling-Nanotube Type of Wiring Transmission Add a New Dimension to Brain Neuro-Glial Networks. Philosophical Transactions of the Royal Society B, 369, Article ID: 20130505.
41. Agnati, L.F., et al. (2010) Understanding Wiring and Volume Transmission. Brain Research Reviews, 64, 137-159. http://dx.doi.org/10.1016/j.brainresrev.2010.03.003
42. Agnati, L.F., et al. (2014) Information Handling by the Brain: Proposal of a New "Paradigm" Involving the Roamer Type of Volume Transmission and the Tunneling Nanotube Type of Wiring Transmission. Journal of Neural Transmission (Vienna), 121, 1431-1449. http://dx.doi.org/10.1007/s00702-014-1240-0
43. Del Giudice, E. and Vitiello, G. (2006) The Role of the Electromagnetic Field in the Formation of Domains in the Process of Symmetry Breaking Phase Transition. Physical Review A, 74, Article ID: 22105. https://arxiv.org/pdf/cond-mat/0607594.pdf. http://dx.doi.org/10.1103/PhysRevA.74.022105

44. Del Giudice, E., Spinetti, P.R. and Tedeschi, A. (2010) Water Dynamics at the Root of Metamorphosis in Living Organisms. Water Journal, 2, 566-586. http://www.mdpi.com/2073-4441/2/3/566. http://dx.doi.org/10.3390/w2030566
45. Montagnier, L., et al. (2015) Transduction of DNA Information through Water and Electromagnetic Waves. Electromagnetic Biology and Medicine, 34, 106-112. https://arxiv.org/pdf/1501.01620v1.pdf. http://dx.doi.org/10.3109/15368378.2015.1036072
46. Jessen, K.R. (2004) Cells in Focus, Glial Cells. The International Journal of Biochemistry & Cell Biology, 36, 1861-1867. http://www.ucl.ac.uk/cdb/research/jessenmirsky/publications/IntJBiochem.pdf
47. Roth, G. and Dicke, U. (2013) Evolution of Nervous Systems and Brains. In: Galizia, C.G. and Lledo, P.M. (eds.) Neurosciences—From Molecule to Behavior: A University Textbook. Chapter 2, Springer-Verlag Berlin Heidelberg, 19-45. http://dx.doi.org/10.1007/978-3-642-10769-6_2
48. Bucher, D. and Anderson, P.A.V. (2015) Evolution of the First Nervous Systems—What Can We Surmise? Journal of Experimental Biology, 218, 501-503. http://jeb.biologists.org/content/218/4/501. http://dx.doi.org/10.1242/jeb.111799
49. Efimov, A., et al. (2007) Asymmetric CLASP-Dependent Nucleation of Noncentrosomal Microtubulcs at thc Trans-Golgi Network. Developmental Cell, 12, 917-930. http://www.cell.com/developmental-cell/abstract/S1534-5807(07)00149-9 http://dx.doi.org/10.1016/j.devcel.2007.04.002
50. Oberheim, N.A., Goldman, S.A. and Nedergaard, M. (2012) Heterogeneity of Astrocytic Form and Function. Methods in Molecular Biology, 814, 23-45. http://dx.doi.org/10.1007/978-1-61779-452-0_3
51. Oberheim, N.A., et al. (2009) Uniquely Hominid Features of Adult Human Astrocytes. The Journal of Neuroscience, 29, 3276-3287. http://dx.doi.org/10.1523/JNEUROSCI.4707-08.2009 www.cns.nyu.edu/events/spf/SPF_papers/Oberheim_et_al_2009.pdf
52. Robertson, J.M. (2013) Astrocytes and the Evolution of the Human Brain. Medical Hypotheses, 82, 236-239. https://www.researchgate.net/publication/259586299_Astrocytes_and_the_evolution_of_

the_human_brain

53. Bondan, E.F., et al. (2015) Glial Cells of the Central Nervous System of Bothrops jararaca (Reptilia, Ofidae): An Ultrastructural Study. Pesquisa Veterinária Brasileira, 35, 685-690. http://www.scielo.br/pdf/pvb/v35n7/1678-5150-pvb-35-07-00685.pdf http://dx.doi.org/10.1590/S0100-736X2015000700014
54. Allen, N.J. and Barres, B.A. (2009) Glia—More than Just Brain Glue. Nature, 457, 675-677. http://dx.doi.org/10.1038/457675a
55. Hartline, D.K. (2011) The Evolutionary Origins of Glia. Glia, 59, 1215-1236. http://dx.doi.org/10.1002/glia.21149
56. Barkovich, A.J., Gressens, P. and Evrard, P. (1992) Formation, Maturation, and Disorders of Brain Neocortex. American Journal of Neuroradiology, 13, 423-446.http://www.ajnr.org/content/13/2/423.full.pdf
57. Lazzari, M. and Franceschini, V. (2004) Glial Fibrillary Acidic Protein and Vimentin Immunoreactivity of Astroglial Cells in the Central Nervous System of the African Lungfish, Protopterus annectens (Dipnoi: Lepidosirenidae). Journal of Morphology, 262, 741-749. https://www.researchgate.net/publication/8229778_Glial_fibrillary_acidic_protein_and_vimentin_immunoreactivity_of_astroglial_cells_in_the_central_nervous_system_of_the_African_lungfishProtopterus_annectens_Dipnoi_Lepidosirenidae. http://dx.doi.org/10.1002/jmor.10274
58. Weissman, T., et al. (2003) Neurogenic Radial Glial Cells in Reptile, Rodent and Human: from Mitosis to Migration. Cerebral Cortex, 13, 550-559. http://cercor.oxfordjournals.org/content/13/6/550.full. http://dx.doi.org/10.1093/cercor/13.6.550
59. Bjornsson, C.S., et al. (2015) It Takes a Village: Constructing the Neurogenic Niche. Development Cell, 32, 435-446. http://dx.doi.org/10.1016/j.devcel.2015.01.010. http://www.cell.com/developmental-cell/pdf/S1534-5807(15)00033-7.pdf
60. Ma, D.K., et al. (2008) Neurogenic Niches in the Adult Mammalian Brain. In: Gage, F.H., Kempermann, G. and Song, H., Eds., Adult Neurogenesis, Chapter 11, Cold Spring Harbor, New York, 207-225.
61. Ming, G. and Song, H. (2011) Adult Neurogenesis in the Mammalian Brain: Significant Answers and Significant Questions. Neuron, 70,

687-702. www.cell.com/neuron/pdf/S0896-6273(11)00348-5.pdf. http://dx.doi.org/10.1016/j.neuron.2011.05.001

62. Morshead, C.M., et al. (1994) Neural Stem Cells in the Adult Mammalian Forebrain: A Relatively Quiescent Subpopulation of Subependymal Cells. Neuron, 13, 1071-1082. http://dx.doi.org/10.1016/0896-6273(94)90046-9
63. Reynolds, B.A. and Weiss, S. (1996) Clonal and Population Analyses Demonstrate That an EGF-Responsive Mammalian Embryonic CNS Precursor Is a Stem Cell. Developmental Biology, 175, 1-13. www.sciencedirect.com/science/article/pii/S0012160696900901. http://dx.doi.org/10.1006/dbio.1996.0090
64. Doetsch, F., Garcia-Verdugo, J.M. and Alvarez-Buylla, A. (1999) Regeneration of a Germinal Layer in the Adult Mammalian Brain. Proceedings of the National Academy of Sciences of the United States of America, 97, 11619-11624. http://dx.doi.org/10.1073/pnas.96.20.11619
65. Oberheim N.A., et al. (2006) Astrocytic Complexity Distinguishes the Human Brain. Trends in Neurosciences, 29, 547-553. http://dx.doi.org/10.1016/j.tins.2006.08.004
66. Fiorelli, R., et al. (2015) Adding a Spatial Dimension to Postnatal Ventricular-Subventricular Zone Neurogenesis. Development (Cambridge, England), 142, 2109-2120. http://dev.biologists.org/content/142/12/2109. http://dx.doi.org/10.1242/dev.119966
67. Maiese, M. (2015) Embodied Selves and Divided Minds—International Perspectives I Philosophy & Psychiatry. Oxford University Press, Oxford.
68. Vigh, B. and Vigh-Teichmann, I. (1998) Actual Problems of the Cerebrospinal Fluid-Contacting Neurons. Microscopy Research and Technique, 41, 57-83. http://dx.doi.org/10.1002/(SICI)1097-0029(19980401)41:1<57::AID-JEMT6>3.0.CO;2-R
69. Vigh, B., et al. (2004) The System of Cerebrospinal Fluid-Contacting Neurons. Its Supposed Role in the Nonsynaptic Signal Transmission of the Brain. Histology and histopathology, 19, 607-628.
70. Tozzi, A. and Peters, J.F. (2015) Brain Activity on Hypersphere. arXiv: 1512.00036 [q-bio.NC]. https://arxiv.org/ftp/arxiv/papers/1512/1512.00036.pdf

71. Peters, J.F., Tozzi, A., Inan, E. and Ramanna, S. (2016) Entropy in Primary Sensory Areas Lower than in Associative Ones: The Brain Lies in Higher Dimensions than the Environment. BioRXiv, 071977. http://biorxiv.org/content/biorxiv/early/2016/08/29/071977.full.pdf
72. 70. Peters, J.F., Inan, E., Tozzi, A. and Ramanna, S. (2016) Primary Evidence of a Donut-Like, Fourth Spatial Dimension in the Brain. BioRXiv, 072397. http://biorxiv.org/content/biorxiv/early/2016/08/30/072397.full.pdf
73. Tozzi, A. (2015) Information Processing in the CNS: A Supramolecular Chemistry? Cognitive Neurodynamics, 9, 463-477. http://dx.doi.org/10.1007/s11571-015-9337-1 http://arturotozzi.webnode.it/products/kindly-helped-by-karl-friston-information-processing-in-the-cns-a-supramolecular-chemistry-/
74. Messori, C. (2004) Le Metamorfosi della Meraviglia. Riflessioni sui Percorsi della Conoscenza dall'Età del Bronzo ad Oggi. Maremmi Editori, Firenze Libri, Firenze.

CHAPTER 2

BIOLOGICAL NEURAL NETWORK STRUCTURE AND SPIKE ACTIVITY PREDICTION BASED ON MULTI-NEURON SPIKE TRAIN DATA

Tielin Zhang[1,2], Yi Zeng[1], Bo Xu[1]

[1]Institute of Automation, Chinese Academy of Sciences, Beijing, China

[2]University of Chinese Academy of Sciences, Beijing, China

ABSTRACT

The micro-scale neural network structure for the brain is essential for the investigation on the brain and mind. Most of the previous studies typically acquired the neural network structure through brain slicing and reconstruction via nanoscale imaging. Nevertheless, this method still cannot scale well, and the observation on the neural activities based on the reconstructed neural network is not possible. Neuron activities are based on the neural network of the brain. In this paper, we propose that multi-neuron spike train data can be used as an alternative source to predict the neural network structure. And two concrete strategies for neural network structure prediction based on such kind of data are introduced, namely, the time-ordered strategy and the spike co-occurrence strategy. The proposed methods can even be applied to in vivo studies since it only requires neural spike activities. Based on the predicted neural network structure and the spreading activation theory, we propose a spike prediction method. For neural network structure reconstruction, the experimental results reveal a significantly improved accuracy compared to previous network reconstruction strategies, such as Cross-correlation, Pearson, and the Spearman method. Experiments on the spikes prediction results show that the proposed spreading activation based strategy is potentially effective for predicting neural spikes in the biological neural net- work. The predictions on the neural network structure and the neuron activities serve as foundations for large scale brain simulation and explorations of human intelligence.

Keywords: Neural Network Structure Prediction, Spike Prediction, Time-Order Strategy, Co-Occurrence Strategy, Spreading Activation

INTRODUCTION

The micro-scale brain anatomy and activities of the neurons are essential for understanding how the brain works and the nature of human intelligence. The general motivation of this paper comes from the fact that neuronal structure and activities are closely connected with each other even though it is usually very hard to get both neural anatomical structure and spike activity by electrophysiology or imaging method at the same time. The general idea is that even without nanoscale imaging techniques (which cannot scale well), it is still possible to reconstruct the neural network structure and predict spike activities by analyzing the spike train data.

From the structure perspective, we propose that multi-neuron spike train data can be used as an alternative source to predict the neural network structures, and here we discuss two concrete strategies for neural network structure prediction based on this kind of data. Multi-neuron spike train data is composed of spike trains from multiple neurons recorded in the same time interval. The data typically shows whether a specific neuron generates a spike or not at a specific time. Two strategies are proposed for neural network structure prediction: 1) the time-ordered strategy: synapses exist between neurons that generate spikes at the two neighborhood time points (i.e. the time point N and N-1); 2) the spike co-occurrence strategy: synapses exist between neurons that fire together at the same time slot. This strategy is consistent with the Hebb's law "cells that fire together, wire together".

From the function perspective, for each neuron in a neural network, its behavior is not only decided by its own property, but also very relevant to its contexts (e.g. other neurons in the same network). Hence, effective prediction of neural spike activities in a network context requires at least the following three major efforts: 1) response prediction of a single neuron towards a stimulus; 2) obtaining the detailed network structure, with synapse information among neurons; 3) modeling signal transmission based on the neural network. We use the neuron simulator to build detailed single pyramidal cells for the first effort. The network structure is built by the methods introduced above, while the spreading activation method based on the two previous efforts is introduced for the third effort. We use part of the spike train data to build the neural network structure and the model for neural activity prediction and the rest is for validation.

In order to validate the proposed methods and strategies, we use the data based on calcium imaging technology. Among different kinds of calcium imaging method, one of them is named functional Multi-neuron Calcium Imaging (fMCI), namely, multi-neurons loading of calcium fluorophores. The advantages of fMCI include: 1) recording en masse from hundreds of neurons in a wide area; 2) single-cell resolution; 3) identifiable location of neurons; and 4) detection of non-active neurons during the observation period. These advantages enables us acquire necessary spike train data discussed above to support the investigation and validation of proposed problems and methods. The paper is structured as follows: Section 2 briefly introduce related work, Section 3 explain the detailed neural network structure and spikes prediction method as well as their relationships, and Section 4 describes the experiments and results to show the prediction accuracy, finally conclusion is made in Section 5. This paper refines and extends the work introduced in [1] and [2].

RELATED WORK

From the structure perspective, most common approaches for obtaining the micro-level neural network structure is typically based on brain slicing and reconstruction with nanoscale imaging [3] [4] . This method can accurately locate the position of neurons, synapses and even dendritic spines. But at the same time, complex and weak nanoscale images make image repairing, synapse recognition and 3D structure reconstruction cost too much time. Based on current brain slicing and imaging techniques, it is still difficult to get synaptic scale neural network structures for a small group of neurons, let alone bigger network. In addition, tissues after slicing are with no functional reactions any more. This makes advanced research on neuronal functions based on the reconstructed neural network nearly impossible.

From the activity perspective at the nanoscale, electrophysiology techniques and calcium imaging techniques are frequently used. The electrophysiology techniques detect and save the neurons' time dependent soma voltage [5] . This technique is with accurate temporal resolution but poor space resolution. In addition, it can only record few neurons' voltage at one time. While calcium fluorescence imaging techniques detect the changes on neurons' voltage values by monitor cell calcium changes, it can monitor the activity of a hundred to thousand neurons simultaneously both in vitro and in vivo [6] . The maximum number of neurons monitored by calcium fluorescence is around 100,000 cells for zebrafish [7] . Stetter uses transfer entropy method to reconstruct the network which requires no prior statistics and connections assumptions, and the work focuses on excitatory synaptic links, and network clustering topology [8] . Takahashi investigates the calcium imaging experiments including imaging, spikes detection and some network analysis tasks. He and coauthors conclude that the network fits for power-law scaling of synchronization properties and network connection probability varies from different network scales [6] .

METHODOLOGY

Although the functional data are collected by functional multi-neuron calcium imaging techniques, the missing of relevant structures from anatomical method still makes analysis about the dynamic causality for the whole network hard. Here we propose architecture to analyze the spike train data to predictively reconstruct the micro- level network structure, and predict neuronal activities based on the reconstructed network.

The methodology introduced here relies on spike train data generated by various kinds of techniques (e.g. fMCI and electrophysiology experiments). In this paper, we use the experimental data from the mouse memory task [6] . Spontaneous spiking activities are still happening in slices of the mouse brain hippocampus CA1 area when it was taken out of the body for not a long time. These spikes show the special task related network function in hippocampus, such as memory retention. The understanding of these kinds of data can greatly speed up the procedure of memory research.

As Figure 1 shows, calcium images and neuron voltages from mouse slices data are collected by fMCI me- thod. Neural voltage signals based on the fMCI technique are with low Signal to Noise Ratio (SNR). Hence, signal processing to filter noise is necessary. Here we use Fast Fourier Transform (FFT) and Inverse Fast Fourier Transform (IFFT) methods to filter the high frequency noises so that spiking signals can be detected precisely. Then two methods to convert neural spike signals to neural structure are introduced, namely, the co-occurrence strategy and the time-ordered strategy. As a step forward, we propose a spike activity prediction method based on the predicted neural network, the spreading activation theory and single pyramidal neuron models.

Voltages Signal Processing

Voltages signals from calcium imaging are so closed combined with noise that high frequency filter must be applied for further spikes detection. The processing procedure mainly includes the following four parts: 1) using FFT to transform time sequence signals to frequency signals; 2) using high frequency filters to eliminate signal noises; 3) using IFFT to transform frequency signals back to time series signals; 4) deciding the voltage threshold to detect spiking signals. As Figure 2 shows, after signal transformation from time series to frequency and back again by FFT and IFFT, high frequency noises are filtered and spike signals are detected.

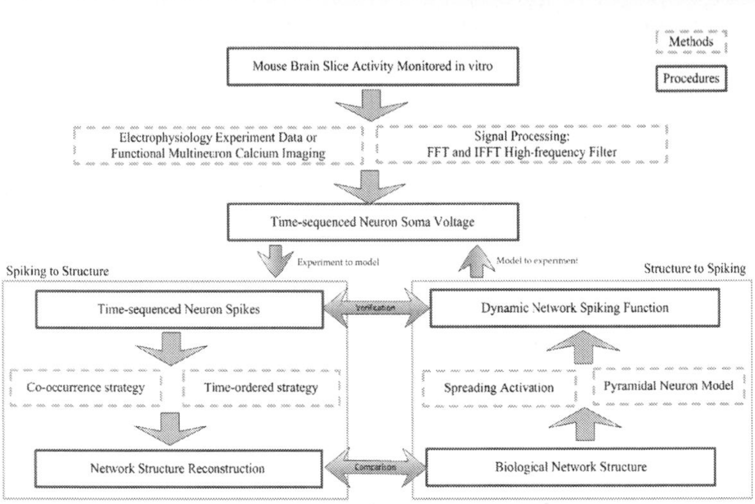

Figure 1. The overall architecture for predicting neural network structure and neuronal activities based on spike train data.

From Spikes to Neuronal Structures

In this paper, two strategies are discussed and used to realize the procedure of converting spike signals to neural network structure:

1) The time-ordered strategy: synapses exist between neurons that generate spikes at the two neighborhood time points (the time point N and N-1);
2) The spike co-occurrence strategy: synapses exist between neurons that fire together at the same time slot. This strategy is consistent with the Hebb's law "cells that fire together, wire together".

Based on the proposed two different strategies, two neurons that may share a connection might be connected from tens to hundreds of times. Each time of connection is with the same unit strength. However, more times of connection between neurons indicate the higher probability for the existence of synapse. As shown in Figure 3, the network is rebuilt by combination of two methods. The upper panel in Figure 3 shows the spikes states of 62 neurons in different time slot. The bottom left panel shows the present spike transfer (in green line). The bottom right panel shows the connection probability of different neurons.

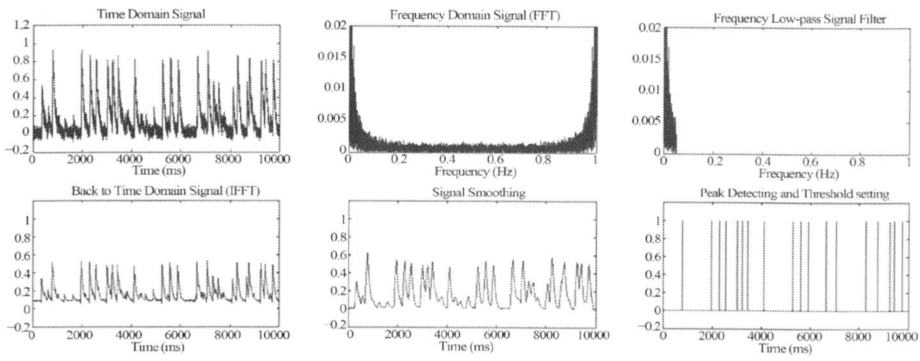

Figure 2. Neural activity signal processing. a) Original time series signals; b) FFT frequency signals; c) High frequency filter; d) IFFTtime series signals; e) Signal smooth process; f) Voltage threshold method of spiking detection.

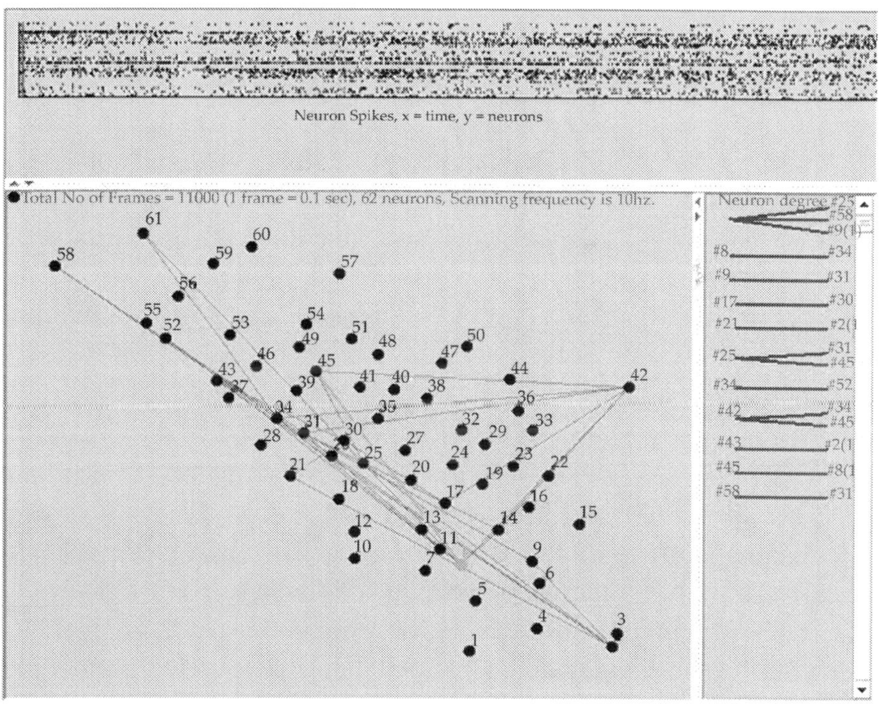

Figure 3. Network structure prediction procedure.

Spike Prediction Based on Spreading Activation

With the reconstructed micro-level neural network based on the methods proposed in Section 3.2, combined with single neuron models, we can predict spike activities for the neurons in observation. Here we first introduce a single neuron model as a foundation. Then we propose a spreading activation based spike prediction method, which utilize the single neuron model and the predicted neural network.

Since pyramidal neuron is widely distributed in mouse hippocampus (more than 90%) [9] , here we introduce the single neuron model of a kind of pyramidal neuron in mouse brain hippocampus CA1 area for the simulation of the neuronal activities. Its neural structure obtained from http://neuromorpho.org/ (ID: NMO_08800) is shown in Figure 4(a). Neuron model contains three basic kinds of ion channels, namely potassium channel, sodium channel and calcium channel. Potassium ion channels mainly include four subtypes, namely K^+-A^+ channel, basic potassium channel, slow Ca^{2+}-dependent potassium channel and calcium-activated potassium channel. The calcium ion channel contains l-calcium channel and basic calcium channel. The sodium ion channel is H-current kind of sodium channel. In addition, detailed ion channel parameters in different model sections such as soma, axon and dendrite are varied from each other. Neuron response with given voltage stimuli is shown in Figure 4(b). Special response from high stimuli voltage can be found in Figure 4(c), which shows the detailed neuron model with more ion channels (not only basic H-H kind of Na^+, K^+, Ca^{2+}), not just leaky integrate-and-fire kind of spiking model. It is able to describe the property of neuronal refractory period well in this model. This detailed pyramidal neuron model built in neuron [10] [11] is essential to understanding and simulation of brain network function.

Spreading activation (SA) was proposed in cognitive psychology for understanding human memory. It is also widely used for searching associative network, neural network or semantic network [12] . The SA network is briefly illustrated in Figure 5. In this paper, this method is applied to the micro scale for understanding neural network activities and predicting spikes.

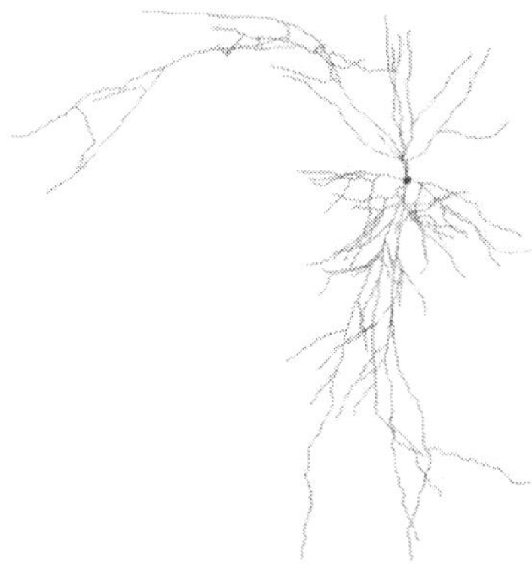

(a) Single pyramidal cell

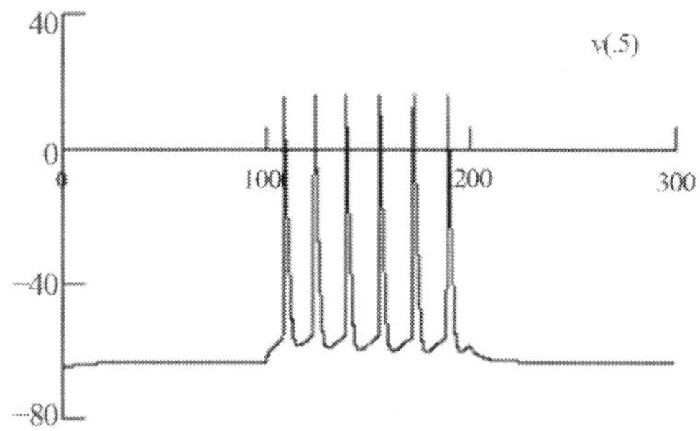

(b) Soma response with 0.1 nA

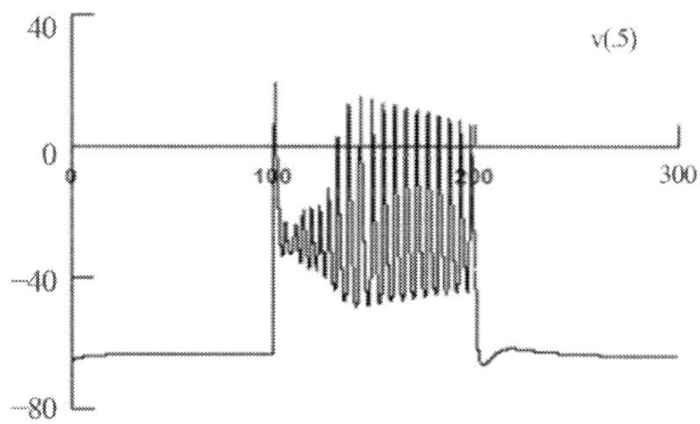

(c) Soma response with 1 nA

Figure 4. Single pyramidal cell's reaction with voltage stimulus (0.1 nA to 1 nA). The morphology of the pyramidal cell (a) is from http://neuromorpho.org/ (ID: NMO_08800).

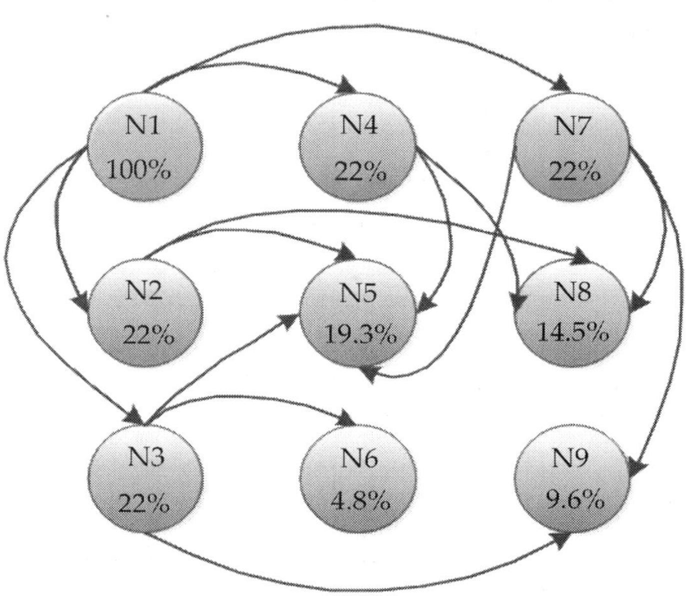

Figure 5. An illustration of the SA network mechanism.

Here we utilize some biology evidence to enhance SA method. The two important variables in SA is the de- cay factor D and the voltage threshold F. As Figure 6 shows, electrical stimulation action experiment is simulated by Neuron software on real pyramidal cell in mouse hippocampus CA1 area introduced above. Both voltages in soma and dendrites are monitored. As Equation (1) shows, since the voltage decay between two neurons is mainly caused by dendrites instead of axons, we can calculate the average voltage decay range as the universal decay which is fit for the decay factor in SA method. Based on the calculation, the value of D should be 0.22. As Equation (2) shows, voltage threshold F can be the neuron spiking threshold which is set as −56 mv based on [13] .

$$D = \frac{\max\left[\frac{\sum_{1}^{N_{dend}} V_{dend} - V_{th}}{N_{dend}}\right]}{\max\left[V_{soma} - V_{th}\right]} \tag{1}$$

$$F = V_{th} \tag{2}$$

D is the decay factor in SA method which can be biologically described as the voltage decay. N_{dend} is the number of neuronal dendrites. V_{dend} is the voltage of the dendrite section. V_{th} is the resting potential of neuron. F, which represents the spreading threshold factor in SA, is described by pyramidal cell voltage threshold.

The SA method is initiated by activating some neurons as source nodes and then iteratively propagates or spreads that activation to other neurons. During the propagation procedure, the values from original neurons de- cay according to different weights of neighbor neurons. The activation will terminate when the values go below the threshold for activation [12] . The detailed SA based method for spike prediction is described as follows.

Spreading Activation Based Method for Spike Prediction:

[1] Initialize the main parameters including voltage threshold F (ranging from zero to one) and decay factor D (ranging from zero to one).

[2] Initialize the network and set all neuron activation value A[i] to zero. Set some origin nodes in A[i] a larger number compared to the voltage threshold F.

[3] For each links [i, j], connect the source [i] with target [j], adjust $A[i] = A[j] + A[i] * W, [i,j] * D$

[4] If a target neuron activation value is larger than threshold F, adjust activation value to Fmax.

[5] Once a neuron has fired, it may not fire again for preventing repeated firing and loops in a refractory period (experimentally set as 10 iteration times).

[6] Neurons whose activation values exceed the voltage threshold F should be marked for firing on the next spreading activation cycle.

[7] Procedure terminates when no more neurons fire or the number of iterations exceeds the maximum value (experimentally set as 10 iteration times).

EXPERIMENTS AND RESULTS

The Pathway Prediction Results

Data of the rat hippocampus CA1 pyramidal cell layer based on fMCI is used for neural network structure prediction (including 8 datasets, and each of them records spike activities for 62 to 226 neurons. The datasets were imaged with the frequency of 10 Hz [6] [14]) using the upper two proposed strategies.

We validate the accuracy of the neural network structure prediction strategies based on the following steps. The overall prediction accuracy is an average value based on the 8 dataset. For each of the dataset: 1) Equally divide one dataset as 20 sub datasets according to the time intervals (The sub datasets are denoted as S1,…,S20). Select the first 80% of the sub datasets (S1,…,S16) for neural network structure construction, and validate the pathway using the rest of the 20% sub dataset (Here we assume that if the predicted pathway is correct, it should cover the neuronal connections based on the rest of the 20% sub datasets); 2) Select another 80% of the sub da- tasets for neural network structure construction, and the rest for validation, and repeat this step until all the sub datasets have been selected for validation. 3) The prediction accuracy is the average value of the 20 predictions.

There are several important observations and indications based on the prediction results. 1) Although the two prediction strategies seem entirely different, the neural network structures based on the two different strategies are very relevant (The correlation is significant with the Pearson correlation value 0.958).

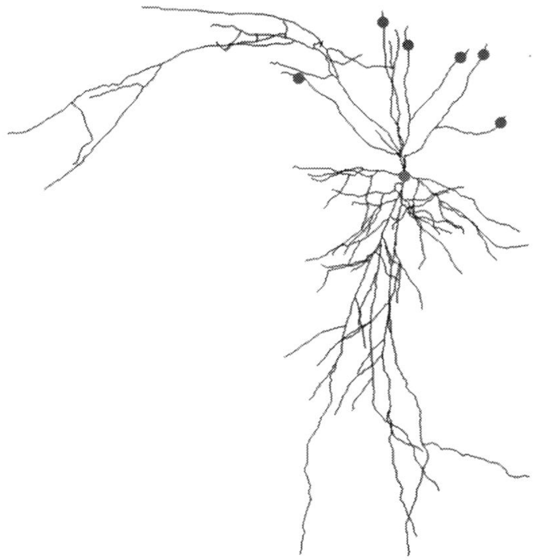

(a)

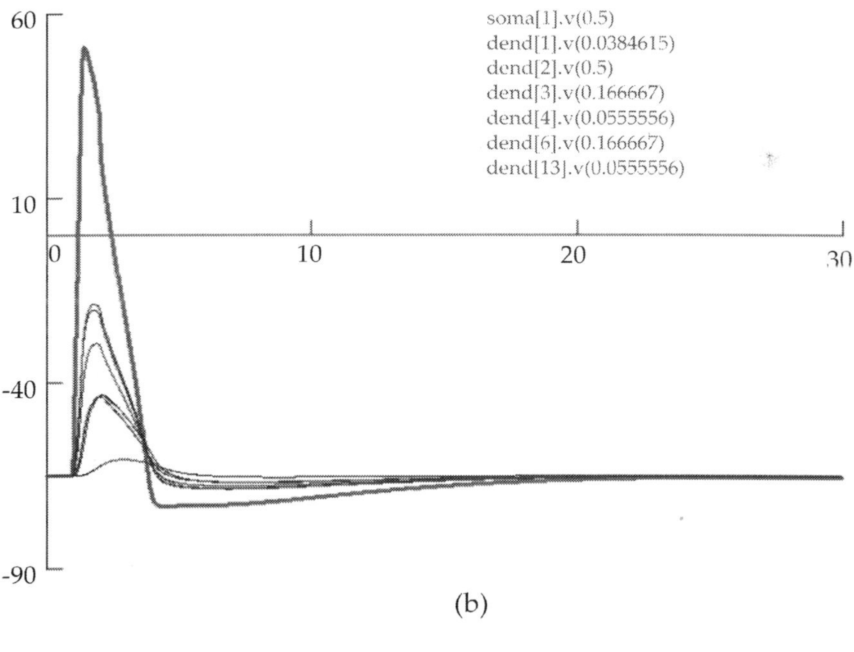

(b)

Figure 6. Pyramidal cell morphology and electrical activity simulation. a) Single pyramidal cell voltage, probe method to ensure the section voltages in soma (red point) and dendrites (blue points); b) Different synapse sections and soma voltage changing by multi stimuli from six neurons. Both soma [i], v[j] and dendrite [i], v[j] sections are built in this model, i stands for the section id, j denotes the detection probe position. The average decay factor is calculated by means of these six voltages.

It indicates that although the proposed strategies are different, the results from the two different strategies do not have major conflicts, instead they are very consistent, and support each other. 2) The neural network structure prediction accuracy for the time-ordered strategy is 83.4%, and the accuracy for the spike co-occurrence strategy is 80.5%. When we group the two neural network structures together (denoted as the merged strategy), the prediction accuracy reaches 89.3%. Figure 7 shows the prediction accuracy for each of the dataset using the proposed strategies. This result indicates that better prediction can be made when the predicted pathways from the two strategies are combined together. 3) 27% of the possible connections among neurons are selected for the time-or- dered strategy, while 25% of the connections are selected by the spike co-occurrence strategy. If the two results are grouped together, 32% of the possible connections are included. The results seem good, since the coverage is not high (and is consistent with the observation by using electro-microscopy techniques [15] [16]), while the predicted accuracy for possible neural network structure reaches 89%. Comparing with other correlation strategies, merged strategy shows better prediction accuracy results as Figure 8 shows.

The proposed method is validated on the data in which the distance of two neurons is within approximately 400 μm [17] . Whether the proposal is applicable when the distance goes further needs to be validated. In addition, our current result is based on fMCI data from rat brain slices. In our future work, we will investigate the possibility of using the proposed method on fMCI in-vivo imaging data.

Spike Prediction Results

Assume a specific neuron (denoted as n1) is connected with N neurons in the network and its action potential is V, the post synaptic neurons of n1 receive transmitted signals from n1. When one synaptic transmission is done and the signal reaches the post synaptic soma, its contribution to this soma is around 5 mv [18] . The overall contribution to the voltage of the soma (denoted as P) is represented as P N = ×5 mv, which is obtained by summing up all the contribution from each of the post synaptic potential, while P is used as the stimulus to generate next action potentials. Each of the potential P for the N neurons that connects to a specific neuron can be obtained through the upper calculation process. Having the structure of the neural network, we predict the neuron which owns the largest value of P will generate a spike in a refractory period (experimentally set as 10 iteration times).

In order to validate the proposed method, the data from the rat hippocampus CA1 pyramidal cell layer using fMCI is used (including 8 datasets, and each of them records spike activities from 62 to 226 neurons. The datasets were pictured with the frequency of 10 Hz [6] [14]). Since the time slot during two neighborhood pictures is 100 ms, signal transmissions may have looped for several rounds. Hence, iterations of the spreading activation process are needed. The spike prediction accuracy for each of the dataset is shown in Figure 9.

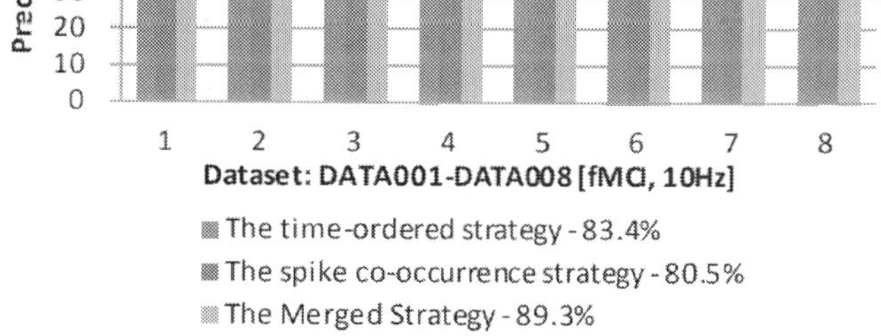

Figure 7. Neural pathway prediction accuracy based on fMCI multi-neuron spike train dataset.

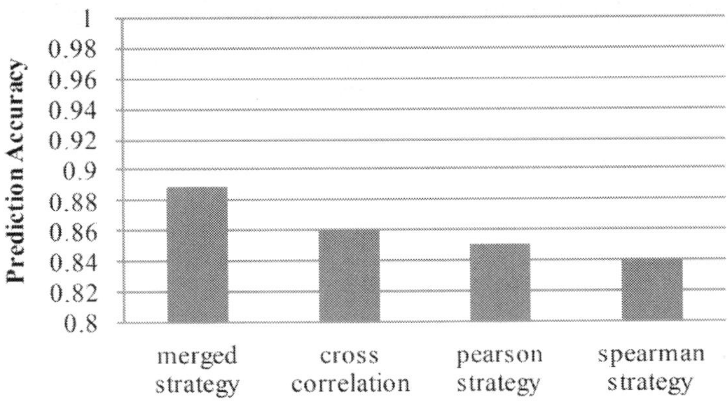

Figure 8. Accuracy made by the merged strategy and other correlation strategies.

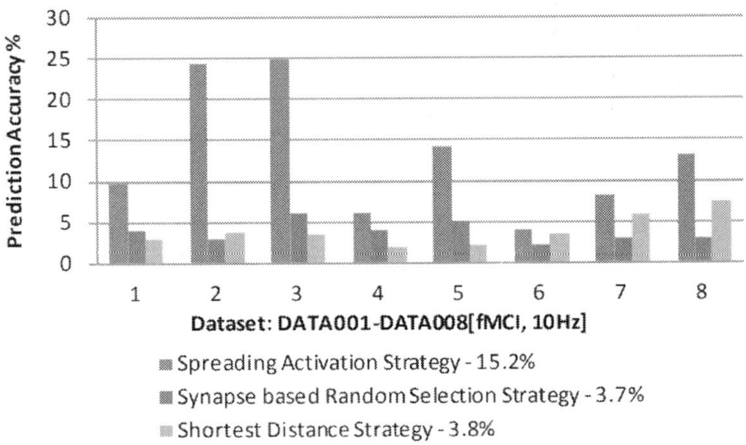

Figure 9. Neural spike prediction accuracy based on different strategies.

As a comparative study, we introduce two alternative strategies, namely the shortest distance strategy (the neuron which owns the shortest distance compared to other post synaptic neurons will be fired), and the synapse based random selection strategy (randomly select a neuron from the set of post synaptic neurons). As shown in Figure 1, the spreading activation based strategy outperforms other two strategies and the average prediction accuracy on 8 datasets is around 15.2%, at least three times better than the other two strategies (the average prediction accuracy for shortest distance strategy is 3.8%, while the synapse based random selection strategy is 3.7%). The validation shows that the proposed spreading activation strategy is potentially effective for predicting neural spikes in the neural network.

CONCLUSION

In this paper, two strategies to predict neural network structure ares proposed and high accuracy experiment results are showed in Section 4.1 and proved their efficiency. This effort makes the rebuilt network structure without anatomy slicing possible. Although restricted by the experimental data, the size of the rebuilt network is around hundred neurons, it still shows great potential for handling larger network reconstruction tasks. Further, the spiking prediction method gives us a view point to analyze the special condition in which we can acquire only structure data. As the first try, this paper only compared with some commonly used methods, such as cross correlation, Pearson and Spearman strategies. More methods will be taken into consideration for comparative studies in the future, such as Granger causality method. One further potential efforts for this work is that spiking functional activities from prediction model can further be used to guide biological experiment signal processing, such as which kind of spikes are noises while others are not.

ACKNOWLEDGEMENTS

Tielin Zhang, Yi Zeng*, Bo Xu This paper is supported by the Brain Engineering project funded by Institute of Automation, Chinese Academy of Sciences. The authors would like to thank anonymous reviewers for constructive comments.

REFERENCES

1. Zeng, Y., Zhang, T.L. and Xu, B. (2014) Neural Pathway Prediction Based on Multi-Neuron Spike Train Data. Proceedings of the 23rd Annual Computational Neuroscience Meeting (CNS 2014), Québec City, 26-31 July 2014, 6.

2. Zhang, T.L., Zeng, Y. and Xu, B. (2014) Neural Spike Prediction Based on Spreading Activation. Proceedings of the 23rd Annual Computational Neuroscience Meeting (CNS 2014), Québec City, 26-31 July 2014, 7.
3. Sporns, O., Tononi, G. and Kötter, R. (2005) The Human Connectome: A Structural Description of the Human Brain. PLoS Computational Biology, 1, e42. http://dx.doi.org/10.1371/journal.pcbi.0010042
4. Arenkiell, B.R. and Ehlers, M.D. (2009) Molecular Genetics and Imaging Technologies for Circuit-Based Neuroanatomy. Nature, 461, 900-907. http://dx.doi.org/10.1038/nature08536
5. Mukamel, R. and Fried, I. (2011) Human Intracranial Recordings and Cognitive Neuroscience. Annual Review of Psychology, 63, 511-537. http://dx.doi.org/10.1146/annurev-psych-120709-145401
6. Takahashi, N., Sasaki, T., Usami, A., Matsuki, N. and Ikegaya, Y. (2007) Watching Neuronal Circuit Dynamics through Functional Multineuron Calcium Imaging (fMCI). Neuroscience Research, 58, 219-225. http://dx.doi.org/10.1016/j.neures.2007.03.001 [Citation Time(s):5]
7. Kettunen, P. (2012) Calcium Imaging in the Zebrafish. In: Islam, S., Ed., Calcium Signaling, Springer Netherlands, Heidelberg, 1039-1071.
8. Stetter, O., Battaglia, D., Soriano, J. and Geisel, T. (2012) Model-Free Reconstruction of Excitatory Neuronal Connectivity from Calcium Imaging Signals. PLoS Computational Biology, 8, e1002653. http://dx.doi.org/10.1371/journal.pcbi.1002653
9. Mira, J. and Sánchez-Andrés, J.V. (1999) Foundations and Tools for Neural Modeling. Proceedings of International Work-Conference on Artificial and Natural Neural Networks, Vol. I, Alicante, 2-4 June 1999, 29.
10. Carnevale, N.T. and Hines, M.L. (2006) The Neuron Book. Cambridge University Press, Cambridge, UK. http://dx.doi.org/10.1017/CBO9780511541612
11. Dayan, P. and Abbott, L.F. (2001) Theoretical Neuroscience: Computational and Mathematical Modeling of Neural Systems.

MIT Press, Cambridge.

12. Anderson, J.R. (1983) A Spreading Activation Theory of Memory. Journal of Verbal Learning and Verbal Behavior, 22, 261-295. http://dx.doi.org/10.1016/S0022-5371(83)90201-3
13. Yue, C.Y., Remy, S., Su, H.L., Beck, H. and Yaari, Y. (2005) Proximal Persistent Na^{+}Channels Drive Spike Afterdepolarization and Associated Bursting in Adult CA1 Pyramidal Cells. The Journal of Neuroscience, 25, 9704-9720. http://dx.doi.org/10.1523/JNEUROSCI.1621-05.2005
14. Yue, C.Y., Remy, S., Su, H.L., Beck, H. and Yaari, Y. (2011) High-Speed Multi-Neuron Calcium Imaging Using Nipkow-Type Confocal Microscopy. Current Protocols in Neuroscience, 2, Unit 2.14. http://www.hippocampus.jp/data
15. Gómez-Di Cesare, C.M., Smith, K.L., Rice, F.L. and Swann, J.W. (1997) Axonal Remodeling during Postnatal Maturation of CA3 Hippocampal Pyramidal Neurons. Journal of Comparative Neurology, 384, 165-180.
16. Fujisawa, S., Matsuki, N. and Ikegaya, Y. (2006) Single Neurons Can Induce Phase Transitions of Cortical Recurrent Networks with Multiple Internal States. Cerebral Cortex, 16, 639-654. http://dx.doi.org/10.1093/cercor/bhj010
17. Sasaki, T., Matsuki, N. and Ikegaya, Y. (2007) Metastability of Active CA3 Networks. The Journal of Neuroscience, 27, 517-528. http://dx.doi.org/10.1523/JNEUROSCI.4514-06.2007
18. Lodish, H., Berk, A., Zipursky, S.L., Matsudaira, P., Baltimore, D. and Darnell, J. (2000) Molecular Cell Biology. 4th Edition, Freeman and Company, New York.

CHAPTER

3

AN EXPERIMENT IN USE OF BRAIN COMPUTER INTERFACES FOR COGNITIVE RESEARCHES

Necmettin Firat Ozkan, Emin Kahya

Department of Industrial Engineering, Eskisehir Osmangazi University, Eskisehir, Turkey

ABSTRACT

Brain-Computer Interfaces (BCIs) are systems that are primarily developed for use of paralyzed people. Although their main aim of use has a medical point of view, they can also be used for different aims such as entertainment and cognitive researches. Since BCI systems have specific brain potentials (P300, steady state evoked potential) and ERD/ERS (Motor Imagery), they are also flexible tools for cognitive science. In this study, an experiment was conducted with 30 participants. Each participant completed two tasks through a BCI and filled NASA-TLX forms. The results were analyzed using paired t-tests to see whether BCI tasks are significantly different in terms of creating cognitive load. The results showed that NASA-TLX scores of the BCI tasks were significantly different and these systems can be considered for estimating cognitive states studies.

Keywords: Cognitive Load, BCI, NASA-TLX

INTRODUCTION

Human error can be defined as some undesired behaviors caused by some improprieties such as lack of knowledge, lack of communication, lack of attention, excessive stress, mental fatigue, and environmental factors. Monitoring cognitive state is a useful tool to reduce these errors. Therefore, many studies about cognitive load have been conducting in several research areas.

According to cognitive load theory, two persons may complete the same task under the same conditions, at the same time and with the same level of success. However, it doesn't guarantee that both people have the same level of cognitive load. It depends on many different factors, such as personal qualifications, attributes, cognitive load, and cognitive state. Although it is important to define the level of cognitive load, there is no available certain metrics to measure. The current methods to measure cognitive load in the literature just try to give some efficient indicators to estimate cognitive load.

Tracy and Albers [1] state that when the cognitive load increases, the ability to perform effectively decreases slowly until the person reaches a point of cognitive overload. According to Hussain et al. [2] , monitoring cognitive load is crucial for developing adaptive systems aware of the user's mental workload. Because such systems can reduce error related risks during task-critical operations. Cain [3] states that a commonly accepted definition of workload does not exist. However, it is possible to characterize workload as a mental structure that reflects the mental strain caused by performing a task under specific environmental and operational conditions, besides the capability of the operator to respond to demands of the task.

Methods for measuring cognitive load can be grouped under three titles: subjective methods, behavioral methods, and physiological methods. Subjective methods are based on self-evaluations of people who completed the relevant tasks. Behavioral methods deal with task completion time and task completion success. These methods usually include secondary task implementations that require completing two tasks concurrently. Physiological methods are used via monitoring some indications from autonomic nerve system and brain activities. Pupil responses, galvanic skin response, and heart rate variability are some of the most common indicators used to estimate cognitive load. EEG and fMRI systems are preferred in studies based on brain activities to estimate cognitive load by monitoring cognitive state.

Antonenko et al. [4] study on comparing NASA-TLX and EEG data and state using EEG is a more objective approach to estimate cognitive load. Haapalainen et al. [5] explain using more than one physical symptom makes the cognitive load estimation more reliable. The study has an experimental part that considers 20 participants. The parameters below were collected from the participants during the experiment:

- Changes on pupil sizes,
- Eye movements,
- Number and duration of eye blinking,
- Heart rate variability,
- Respiration, and
- Monitoring brainwaves.

Walczyk et al. [6] explain that when somebody is telling a lie, his/her pupil sized shows some changes. They conducted a scenario based experiment with 145 participants to identify pupil size differences between telling truth and lie.

BCIs mainly have medical aims but they also have huge potential to be used in different areas of life. Since they are being activated via specific brain activities brain based cognitive researches. They usually offer three kinds of applications that are activated through P300 potential, steady state visual evoked potential, and event related desynchronization/synchronization. In this study, one of the recently developed BCI system's letter matrix applications was utilized. Different difficulty levels and cognitive load of participants was measured via a well-known subjective method: NASA-TLX.

NASA-TLX

Since subjective methods are easy to use, low cost and provide simple quantitative results, they still keep their popularity. One of the widely used subjective methods is NASA-TLX. This method was developed in Hart and Staveland [7] and its application areas were investigated by Hart [8] after 20 years development. There are many other numbers of studies that used NASA-TLX in the relevant literature. Rubio et al. [9] conducted a comparison with NASA-TLX and another subjective method SWAT. Noyes and Bruneau [10] searched differences between use of pencil and computer versions of NASA-TLX. Alm and Nilsson [11] investigated negative effects of using mobile telephone during driving via NASA-TLX method. Yurko et al. [12] stated that higher mental workload caused poorer laparoscopic performance. A paper published by Miyake [13] offered a combined workload estimating method including physiological and subjective methods.

NASA-TLX has six sub-scales which reflect participant's feelings and opinions about the task. The sub-scales are listed below:

- Mental demand,
- Physical demand,
- Temporal demand,
- Performancc,
- Effort,
- Frustration.

Participants evaluate the task with these scales from 0 to 100. It is also possible to decide weights of the scales with participants' evaluations. Fifteen paired comparisons are available and the participant decides more considerable scales for each comparison. Finally it provides an overall score between 0 and 100 to evaluate the difficulty level of the task.

Computer Interfaces

BCIs are technological systems including a computer, relevant software, EEG components (bonnet, electrodes, etc.), and an amplifier. The system works based on EEG and simply brain activities by converting them to comments for the computer using interfaces [14] . Figure 1 illustrates working steps of BCI systems. The main goal of BCI research for the last thirty years is to help paralyzed people for their daily life. However, it is also possible to use these interfaces for other cognitive researches related with ERD and other brain activities.

Karagoz et al. [15] state that BCI systems utilize three main neurophysiological facts. One of them is P300 potential that appears when a problem is solved, a decision is made, or an fogginess disappears in 300 ms. Alkaç [16] stated that P300 doesn't appear with the standard stimulates. It requires specific target stimulates. That's why this potential is used in the studies about gathering attention and focusing. The relevant interface of P300 is usually a letter matrix that columns and rows flash and the participant tries to focus on his/her target letter to type it on the screen.

Since P300 potential has a close relationship with decision making, problem solving and focusing on target stimulus, it is also an attractive brain fact for cognitive load researchers. Schultheis and Jameson [17] designed an experiment with a task including reading written materials with various level of difficulty. They used an EEG, remote eye tracker, and subjective scale to measure the cognitive load. Based on the results, it can be satated that more difficult texts led to lower reading speed, higher subjective load ratings, and a reduced P300 amplitude.

SSVEP is an another fact that is being used by BCI systems. BCIs generate this potential with flashing controls. The direction indicators flash with various frequencies and the participant tries to focus on the target indicator to give the comment. This application makes possible to control some electronic systems [18] . Also, It requires a high level of attention and concentration to be able to catch quick flickers. Further, SSVEP is an useful fact for cognitive load researchers. Perego et al. [19] conducted an experimental study and they stated that the test showed SSVEP based BCI can be used for psychometric test.

The third neurophysiologic fact used by BCIs is changes on motor zones of the brain when participant focuses on thinking amovement to a specific direction [18] . When we think to move our right hand, an activity appears on the left side of our brain. BCI system converts it to a comment for the computer to move cursor on the screen.

METHOD

In this study, an experiment was designed using a BCI system developed by G-TEC Company. The system includes a letter matrix (Figure 2) to type letters via P300 potential.

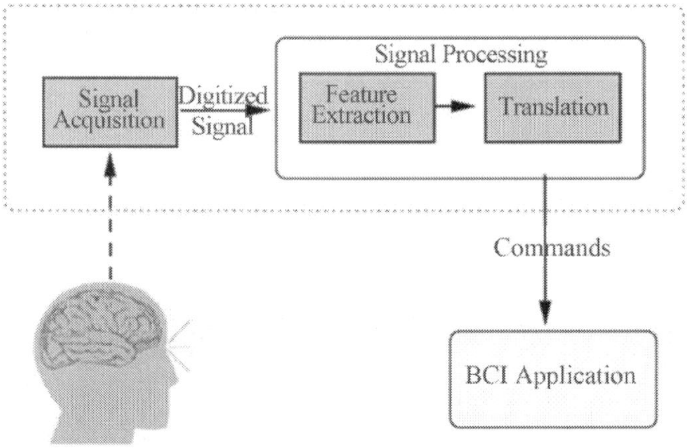

Figure 1. Operation Diagram of BCI [20] .

Thirty 18 - 27 aged participants (15 female,15 male) were recruited for this experiment. Each participant completed two tasks using the letter matrix. All the experiments were conducted in the same silent room with same climatic, illumination conditions. For running BCI, the same computer (Core i5, 8 GB Ram, 15.6 inch screen) was used.

Before executing the task EEG, electrodes were mounted. Figure 3 illustrates the locations for electrode placement for using the matrix.

Task 1: The first task was defined as typing "ABCDE" on the screen. Letter matrix's rows and columns started flashing and participants must have focused on the target letters one by one starting from the first letter. Each target letter flashed 15 times for 75 ms. Besides each target letter had flashed for a second right before the rows and columns' flashes started to make the participant realize its position on the matrix.

Task 2: After Task 1 had been completed, participants filled a paper version NASA-TLX form and took a rest for 5 minutes. For Task 2, participants were asked to type "ESOGU2014" on the screen. For this task, consecutive target letters were not assigned side to side. There was no flashing to show letter's initial position. Therefore, participants had to find the letter's location by themselves during 2 seconds that is a constant time period between typing a letter and starting flashing for the next one. In this task, each target letter flashed 5 times for 50 ms. After completing Task 2, participants filled NASA-TLX form again. It was expected that Task 2's performances would be lower and mental workload score would be higher.

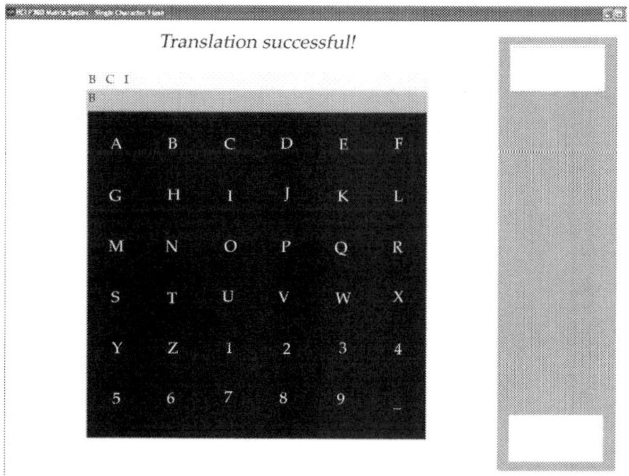

Figure 2. P300 Letter Matrix.

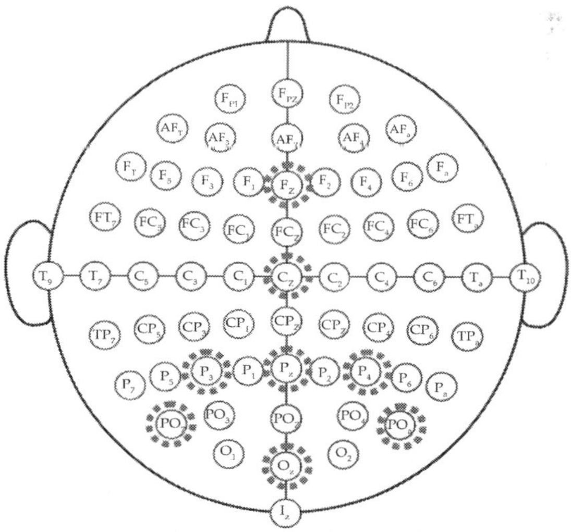

Figure 3. P300 Electrode Placement Locations.

RESULTS

All participants' task performances were recorded (number of correct letters/total number of letters) and NASA- TLX scores were calculated for both tasks. Since Task 2 was more difficult, it was expected that NASA-TLX scores for Task 2 would be dramatically lower. Appendix A and Appendix B illustrate scale weights and overall NASA-TLX scores and Figure 4 summarizes task performances of all participants.

Due to longer word, no help for finding letter on the matrix, quicker, and shorter flashings a huge decrease is recorded in Task 2 performances. Average success levels are 64.00 and 34, 81 for task 1 and task 2 respectively.

As seen in Appendices A and B, better overall NASA-TLX scores was obtained for task 1. Difference between difficulty levels of the tasks can be seen from the average values. The most dramatic change was provided for mental demand in terms of raw scores. When the weight of the scales is investigated, the largest change belongs to weight of temporal demand due to quicker flashings and trying to find the next target letter during Task 2.

ANALYSIS

A paired t-test was conducted to assess whether the difference between overall NASA-TLX scores is significant or not. Before conducting paired t-test, normality of the overall scores was checked with Anderson-Darling test in Minitab 16. Both tasks' NASA-TLX scores distributed normally (Figure 5and Figure 6).

Result of paired t-test is presented below. Since P-value is less than 0.05, we can say that mental workload levels of the tasks are statistically significant (Table 1).

CONCLUSION AND FUTURE WORK

In this study, it aims to see how possible it is to use BCI systems for cognitive load measurement and cognitive state monitoring. Results show that it is possible to use these systems with at least one brain potential including well-determined task parameters.

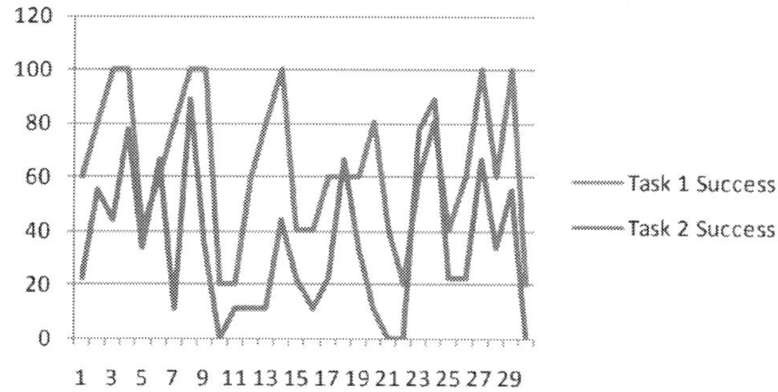

Figure 4. Task Performances.

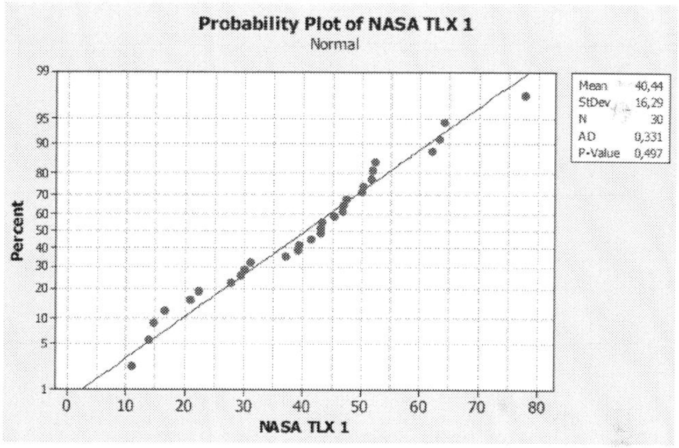

Figure 5. Normality Test of Overall NASA-TLX Score for Task 1.

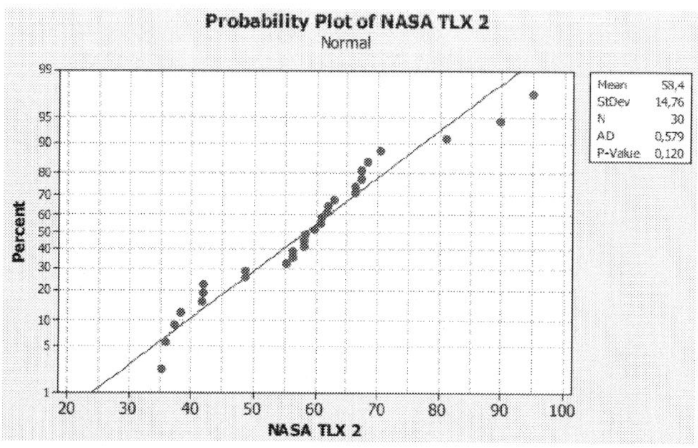

Figure 6. Normality Test of Overall NASA-TLX Score for Task 2.

Table 1. Paired t-test result for overall NASA-TLX scores.

	N	Mean	StDev	SE Mean
NASA-TLX 1	30	40.44	16.29	2.97
NASA-TLX 2	30	58.40	14.76	2.69
Difference	30	−17.96	19.87	3.63

95% CI for mean difference: (−25, 38; −10, 54); t-test of mean difference = (vs not = 0); T-value = −4.95; P-value = 0.000.

Obtaining significant differences in terms of subjective mental load assessment indicates that these systems can be used as a member of combined methods of estimating cognitive state or measuring cognitive load. With the support of some physiological parameters, these systems can become a wide task designing and application platforms. In the future studies, this BCI system will be combined with a remote eye tracker and Galvanic Skin Response device and a new experiment can be designed with more participants.

ACKNOWLEDGEMENTS

Necmettin Firat Ozkan, Emin Kahya (No. 201315A208) which is supported by Eskisehir Osmangazi University Scientific Research Projects Office.

REFERENCES

1. Tracy, J.P. and Albers, M.J. (2006) Measuring Cognitive Load to Test the Usability of Web Sites. Society for Technical Communication 53rd Annual Conference, Las Vegas, 7-10 May 2006, 256-260.
2. Hussain, S., Chen, S.Y., Calvo, R.A. and Chen, F. (2011) Classification of Cognitive Load from Task Performance & Multichannel Physiology during Affective Changes. MMCogEmS: Inferring Cognitive and Emotional States from Multimodal Measures, ICMI 2011 Workshop, 17 November 2011, Alicante.
3. Cain, B. (2007) A Review of the Mental Workload Literature. Defence Research and Development Canada, Toronto.

4. Antonenko, P., Paas, F., Grabner, R. and van Gog, T. (2010) Using Electroencephalography to Measure Cognitive Load. Educational Psychology Review, 22, 425-438. http://dx.doi.org/10.1007/s10648-010-9130-y
5. Haapalainen, E., Kim, S., Forlizzi, J.F. and Dey, A.K. (2010) Psycho-Physiological Measures for Assessing Cognitive Load. Proceedings of the 12th ACM International Conference on Ubiquitous Computing, Copenhagen, 26-29 Septem- ber 2010, 301-310. http://www.ubicomp.org/ubicomp2010/
6. Walczyk, J.J., Griffith, D.A., Yates, R., Visconte, S.R., Simoneaux, B. and Harris, L.L. (2012) Lie Detection by Inducing Cognitive Load Eye Movements and Other Cues to the False Answers of "Witnesses" to Crimes. Criminal Justice and Behavior, 39, 887-909. http://dx.doi.org/10.1177/0093854812437014
7. Hart, S.G. and Staveland, L.E. (1988) Development of NASA-TLX (Task Load Index): Results of Empirical and Theo- retical Research. Advances in Psychology, 52, 139-183. http://dx.doi.org/10.1016/S0166-4115(08)62386-9
8. Hart, S.G. (2006) NASA-Task Load Index (NASA-TLX); 20 Years Later. Proceedings of the Human Factors and Er- gonomics Society Annual Meeting, 50, 904-908.
9. Rubio, S., Díaz, E., Martín, J. and Puente, J.M. (2004) Evaluation of Subjective Mental Workload: A Comparison of SWAT, NASA-TLX, and Workload Profile Methods. Applied Psychology, 53, 61-86. http://dx.doi.org/10.1111/j.1464-0597.2004.00161.x
10. Noyes, J.M. and Bruneau, D.P. (2007) A Self-Analysis of the NASA-TLX Workload Measure. Ergonomics, 50, 514- 519. http://dx.doi.org/10.1080/00140130701235232
11. Alm, H. and Nilsson, L. (1995) The Effects of a Mobile Telephone Task on Driver Behaviour in a Car Following Situation. Accident Analysis & Prevention, 27, 707-715. http://dx.doi.org/10.1016/0001-4575(95)00026-V
12. Yurko, Y.Y., Scerbo, M.W., Prabhu, A.S., Acker, C.E. and Stefanidis, D. (2010) Higher Mental Workload Is Associated with Poorer Laparoscopic Performance as Measured by the NASA-TLX Tool.

Simulation in Healthcare, 5, 267-271. http://dx.doi.org/10.1097/SIH.0b013e3181e3f329

13. Miyake, S. (2001) Multivariate Workload Evaluation Combining Physiological and Subjective Measures. International Journal of Psychophysiology, 40, 233-238. http://dx.doi.org/10.1016/S0167-8760(00)00191-4
14. Ozkan, N.F. and Kahya, E. (2014) Measurement of Cognitive Workload by Use of Combined Methods Including Brain-Computer Interfaces. Advances in Social and Organizational Factors, 12, 458-468.
15. Karagöz, M., Alkaç, U., Ergen, N., Erdamlar, N. and Alpkan, L. (2005) Psikiyatrik Hastaliklarda Elektrofizyolojik (P300) Yöntemler. Düşünen Adam, 18, 210-216.
16. Alkaç, U. (2009) Beyin Araştirmalari Tarihinde Bir Gezinti: Elektronörofizyoloji. Klinik Gelişim, 3, 14-19
17. Schultheis, H. and Jameson, A. (2004) Assessing Cognitive Load in Adaptive Hypermedia Systems: Physiological and Behavioral Methods. In: De Bra, P.M.E. and Nejdl, W., Eds., Adaptive Hypermedia and Adaptive Web-Based Systems, Springer Berlin, Heidelberg, 225-234. http://dx.doi.org/10.1007/978-3-540-27780-4_26
18. Gencer, N., Usakli, A.B., Erdogan, H.B., Akinci, B. and Acar, E. (2010) ODTÜ Beyin Araştirmalari Laboratuvari: Beyin-Bilgisayar Arayüzü çalişmalari. ODTÜLÜ, 45, 20-21 [Citation Time(s):2]
19. Perego, P., Turconi, A.C., Andreoni, G., Maggi, L., Beretta, E., Parini, S. and Gagliardi, C. (2011) Cognitive Ability Assessment by Brain-Computer Interface: Validation of a New Assessment Method for Cognitive Abilities. Journal of Neuroscience Methods, 201, 239-250. http://dx.doi.org/10.1016/j.jneumeth.2011.06.025
20. www.wired.com.

CHAPTER 4

A CHESSBOARD MODEL OF HUMAN BRAIN AND AN APPLICATION ON MEMORY CAPACITY

Chenxia Gu, Shaotong Wang, Hao Yu

Department of Mathematical Sciences, Xi'an Jiaotong-Liverpool University, Suzhou, China

ABSTRACT

The famous claim that we only use about 10% of the brain capacity has recently been challenged. Researchers argue that we are likely to use the whole brain, against the 10% claim. Some evidence and results from relevant studies and experiments related to memory in the field of neuroscience lead to the conclusion that if the rest 90% of the brain is not used, then many neural pathways will degenerate. What is memory? How does the brain function? What would be the limit of memory capacity? This article provides a model established upon the physiological and neurological characteristics of the human brain, which can give some theoretical support and scientific explanation to explain some phenomena. It may not only have theoretically significance in neuroscience, but can also be practically useful to fill in the gap between the natural and machine intelligence.

Keywords: Memory Capacity, Excitation Transmission, Neuroscience, Machine Intelligence

INTRODUCTION

The memory of a human being is far more complex than storing and retrieving stored information in a computer. It includes the association and inference [1] [2] among information, e.g., the association between two pieces of information such as the appearance of an apple and the human-language name, apple. To store the association, neural pathways will be generated, i.e., connections will be made and more neuron cells must be involved than to simply memorize the two pieces of information [3] . The nerve cells can perform two states excitatory and inhibitory. The transformation of a nerve cell between the two states depends on the irritation obtained from the external environment [4] [5] . The memory forms and stores in hippocampus which consists of a large number of nerve cells. Every pair of nerve cells can generate a connection named synapse which is a channel for transmitting the nervous impulse from one cell to another. In addition, synapse can only conduct impulse in one direction, which means impulse cannot transmit back to the original cell through the same synapse [6] . In terms of the formation delivered by synapse, the excitatory nerve cells have a probability to establish synapse with adjacent cells. The excitement tends to propagate from the cells connected with excitatory sensor cells to the target neuron cells. Whenever the neuron cell state changes, the appeared synapse can always be used as a channel for conducting excitement. The pattern of connections among cells is regarded as the memory of creatures [7] .

Inspired by biological neural networks, artificial neural networks (ANNs) are a family of statistical learning models used to estimate or approximate unknown functions that can depend on a large number of inputs, which are presented as systems of interconnected "neurons" which exchange messages between each other [1] . Inheriting the structure of ANNs and its application on object recognition, a model is established to explain the functioning of the brain, based on the physiological and neurological characteristics of the human brain. Binary number will be used to describe the state of one neuron cell where 1 means excitatory and 0 means inhibitory. The two states of a cell can change when the condition is just right [8] . Then the bridges of each pair of nerve cells represent the synapse. Finally, the outermost layer of the processor matrix represents the target cells which are the end of excitement propagation [9] .

MODELLING

The model consists of a sensor-matrix layer and a processor-matrices layer, each of which contains a certain number of sensor-matrices and processor-matrices respectively. The reason why we can work within two-dimension matrix is that the exact positions of neuron cells are not so important, since essentially speaking, memory depends on not the neurons themselves, but the connections between them. Moreover, due to the complexity of the real process in the human brain which could be far beyond our imagination, not even within three dimensions could it be fully illustrated [10] - [12] .

Sensors are used to generate perception which responds to a stimulation from the outside world. For instance, our eyes, as a visual sensor, can sense the visual features of the stimulation; ears can sense vocal information and hands can sense tactual information. All the information will then be processed and transmitted to the processors, where memory-making process takes place (hippocampus) [13] . After the processors received the information from the sensors, the different sorts of information will be associated together in some ways and forms an initial acquisition of the memory subject. The whole memory process involves three steps, Encoding, Consolidation and Storage, and Retrieval [14] [15] . The transition process from sensors to processors is called encoding step. The association of all sorts of information that the processors received from sensors is regarded as the step for consolidation and storage. For instance, if the visual features of an apple and the literal information of an apple are associated in a memory process, then next time, one could recall the name, apple, as soon as he or she see an apple. The recalling is viewed related to the retrieval step.

Although this model does not require a particular memory representation, it is useful to illustrate the realization of all the three steps with a simplified representation.

Encoding

To simulate transition process, the system with sensors and processors is considered. Each state of the sensors or processors can be stored in a matrix, where each entry represents the state of each neuron cell. The state of a neuron cell is assumed to be either excitatory or inhibitory, which is indicated by binary numbers, that is, 1 and 0, respectively. An example of the encoding process from the sensors to the processors is illustrated in Figure 1.

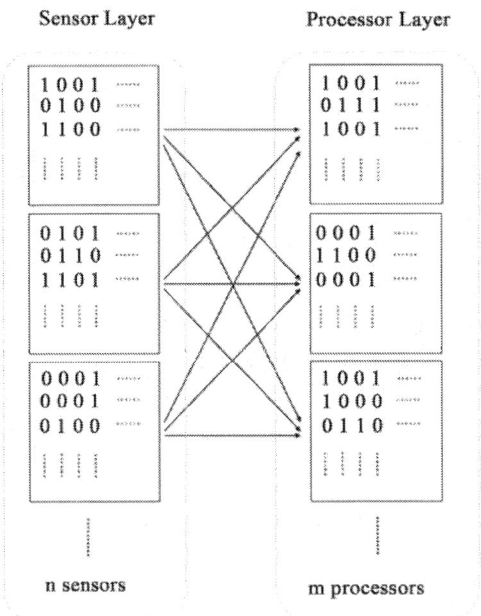

Figure 1. Into m correspondence: an excitatory cell is denoted by the number 1.

Every cell in the sensor matrix could have its own multiple corresponding cells in the processor matrix. If one cell in the sensor matrix is set to be 1, then each of the corresponding cells in the processor matrix will have the value 1 as well [16] . In this model, the correspondence is set to be one-to-one and is randomly generated. The randomness of correspondence is of computational significance, since it disperses the distribution of the cells with value of 1 in the processor which maximizes the utilization of the matrix and maximizes the distinctions of the information in the sensor. Imagine if the cells with value 1 in the processor are all concentrating in the center, the results from the operation of the model will be hard to distinguish.

Consolidation and Storage

Now constrain our attention to one of the sensors and its corresponding processor, say visual sensor and its corresponding processor, and the correspondence between them is assumed to be one-to-one. Therefore, the visual processor, which has a spatial structure in the brain, could be simply represented by another n by n matrix. Because the message delivered fast in human brain, the all duration of the transmission from first cell to another cell are regarded equally [17] . At first, the processor received an initial state of excitation from the sensor. Then, the excitation transmission process starts. (It is assumed that the excitation could only conduct from one cell to another once in one stage of the process and it could only be conducted from the newly excited cells. In another word, the cells that have successfully conducted excitation will not be able to conduct excitation any more. After the transmission from first cell to another, a new state of this matrix will be generated). Before the conduction of excitation, the path which used to transmit the excitation between a pair of cell is generated firstly. When the conduction process is finished, the path will be remained there, which forms the memory. The term bridge is used to indicate whether there is path from one cell to another, and noticed that every bridge is represented by a directed line segment. If there is a bridge from cell A to cell B, then it means excitation can be conducted from cell A to cell B. A pair of cells could have at most two bridges between them, opposite in direction. For an illustration, the cues of the n by n processor matrix are assumed to lie on the four edges excluding the points at the corners, as is shown in Figure 2. The inner matrix represents the processor with some initial state and the red cells on the peripheral edges around this matrix represent the cues.

The conduction process in a processor is regarded as finished once all the cues received the excitation transmitted by bridge and series of excitation cell. The transmission of the excitation is under some conduction rules with neurological significance and it is obvious that the pathway of the excitation will be governed by the cues. Once the initial configuration and the cues are determined, the excitation will be under conduction between cells in the processor and ends up with its reach to the cues. One possible pathway is also shown in Figure 2 where the bridges are indicated by the blue line segments; red cells represent cues.

Conduction Rules

To deduct the possible pathway, conduction rules are established to ensure that all the cues could be reached and the establishment of the bridge is optimized. For better illustration, several rules have been declared. Firstly, the conduction only occurs when bridge has been established between the two cells. Secondly, design that each cell can build up bridges to its four adjacent cells, with directions left, right, up and down respectively.

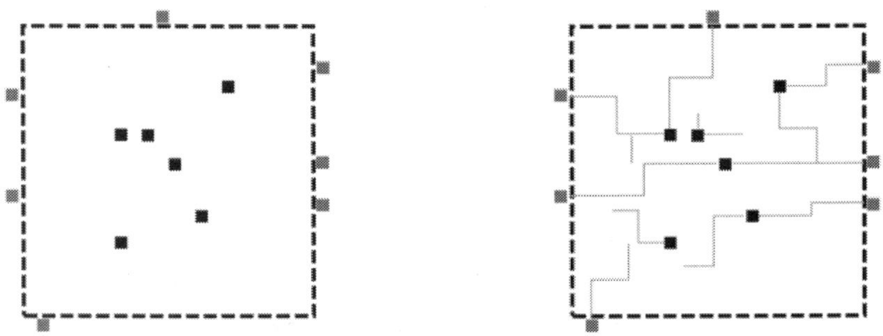

Figure 2. Initial configuration and one possible pathway.

The circumstance with more connections will be considered in later analysis. Thirdly, the transmission process is assumed to be inherently probabilistic. The distance to a cue determines the probability whether the bridge will be established between the cell and the next cell which is nearer to that cue.

Let O and I denote the sets of cues and original excited cell respectively. Suppose that there are totally m cues and n original excited cells. That is to say, $O = O_1, O_2, O_3, \dots, O_m$ and $I = I_1, I_2, I_3, \dots I_n$. Then theoretically, the conditional $f_{O_i I_3} = prob(O_i|I_j)$ is determined by

$$f_{O_iI_j} = \frac{P_{O_iI_j}}{\sum_{k\in O} P_{O_kI_j}} \quad (1)$$

where P denotes $prob(O_r \cup I_t)$, the probability that events O_r and I_t happen simultaneously. Obviously,

by assumption, the instantaneous set of unreached cues governs the conduction of the next moment. Moreover, the cues could have accumulative effects on the probability of the bridge establishment. After the transmission process, the bridges are kept and stored in the processor, which is regarded as memory.

In the second time, when the figure of apple irritates the sensor again, the excitement will be transmitted automatically through the path established in the first time and all cues will be irritated again which indicates the name, apple.

Illustration

The model is aimed to simulate the memory process and explore the memory capacity. In another word, it is aimed to find out the limit beyond which the brain would be overloaded and the association of different types of information could be unreachable. For better illustrate, the model is now applied to a specific example with a 30*30 processor matrix. The initial configuration is initialized as shown in Figure 3.

The black dots inside represents the initial state of the processor, which could be an image of an apple. If the red dots outermost refer to the name, apple, then after application of the model, bridges will be established through which the excitation will be conducted until the reach of excitation to all the cues eventually. One possible bridge is shown in Figure 4.

After that, the bridges will be kept as a memory which associates the vision of the apple and its name. Then next time, if the processor receives the information of the apple again, the excitation will be transmitted along the bridges stored automatically and finally reach the cues which indicates the name of the apple. If the processor receives a different piece of information, the excitation will still be transmitted but may not be associated with the same cues, the name apple. For instance, if an image and its name (cues represented by red cells) are defined as shown in Figure 5(a), another different image is indicated in Figure 5(b) input in a short time. Then the name output in Figure 5(a) is different from it in Figure 5(b), so this is a successful case. However in Figure 5(c), another different image obtained by processor in a short time, the name output is totally the same as it in Figure 5(a), which means the images are not be distinguished. Therefore this is a chaotic case [18] .

RESULTS AND ANALYSIS

Whether the system is chaotic depends on the state of the cells [19] . In order to describe the chaotic degree, the concept similarity is introduced here. Two pieces of different image information are delivered to the same processor in a short period, which means when the second image message is obtained; the bridges established for the first image have not degraded. Two cues, one is for the first message and the other is for the second message.

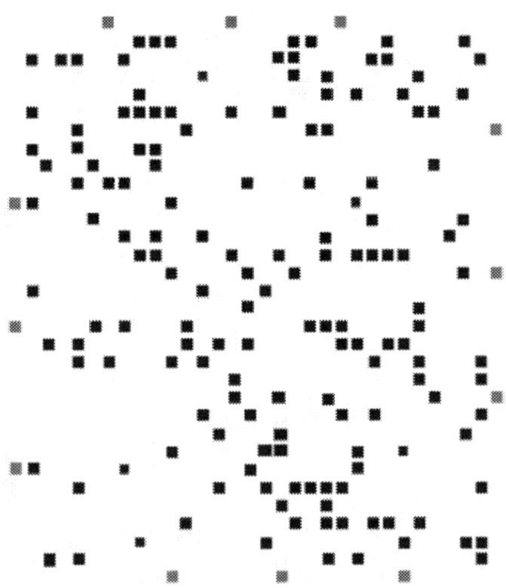

Figure 3. Initial configuration with black dots indicating the cells in the processor and red dots indicating the cues.

Figure 4. One possible bridge.

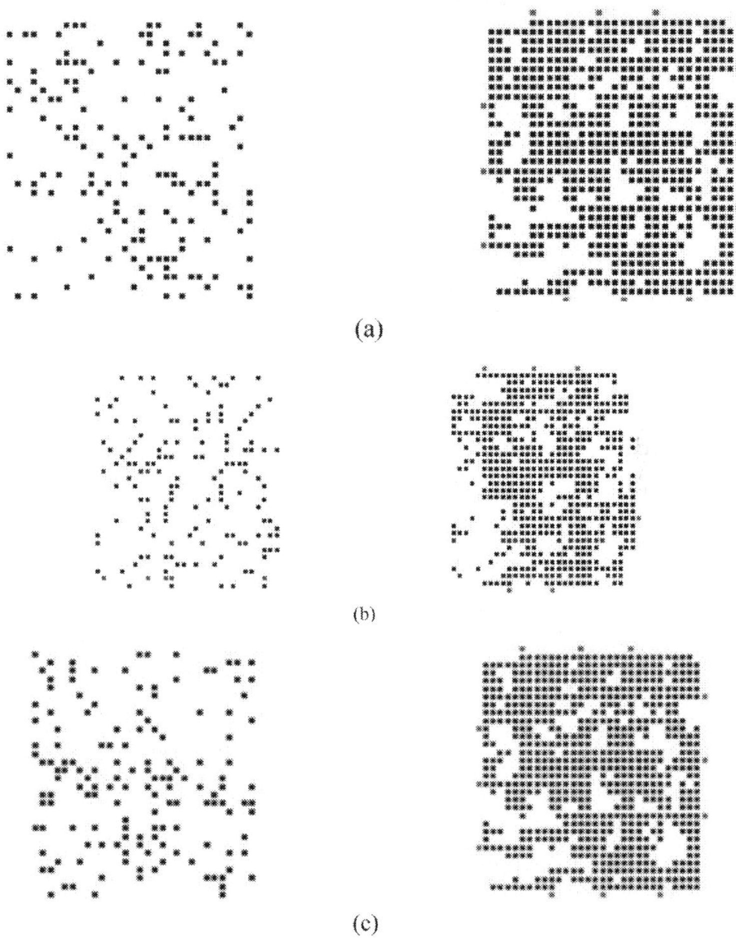

Figure 5. (a) The reference image and its name; (b) Successfully-Distinguished Case: the processor successfully distinguishes a different image; (c) Chaotic

Case: the associated cues are exactly the same as shown in (a), while the information is quite different.

Some parts in the two cue-layers generated respectively are supposed to be different, and the percentage of the same part in the two cue-layers is indicated by similarity. However if the two cue-layers are exactly the same, the case is regarded as a chaotic case. First of all, assume that the number of cells with value of 1 in inner matrix represents the number of information cells included in the image. The relationship between whether the processor is chaotic and the number of information cells involved is investigated by the following method. First, change the percentage of the number of cell with the value of 1 among total number of cells both in image matrix and cue layer from 1% to 90%. Then, repeat each test 20 times for each percentage and record the similarity of each different test. The similarity (S) of the results can be calculated by:

$$G_{ij} = \begin{cases} 0 & \text{if } i \neq j \\ 1 & \text{if } i = j \end{cases}, \quad S = \frac{\sum_{t=1}^{n} G_{a_t b_t}}{n} \times 100\% \tag{2}$$

where n is the total number of bits in cue layer; a_t and b_t represent the tth bit in the test sample and the original sample, respectively.

When S equals 1, which means the model cannot tell the difference between the two different images, which is called chaotic. If G is 0, that means the two relative cues are not exactly the same, the difference of the two different images is told by the model. In order to find an optimal condition to distinguish the different image information, first of all, assume that the number of cells with value of 1 in image matrix represents the amount of information stored in the image matrix. The relationship between the whether the processor is chaotic and the number of cells is investigated by using the following method. Adjust the percentage of the cells storing information and located in image matrix and cue layer from 1% to 90%. Repeat 50 times in every percentage and record the times where the similarity is not equal to 1. Then calculate the successful rate under each condition, where the successful rate is the percentage of the experiment with S = 1 in the whole 50 times experiment. Therefore, Table 1 shows the results.

Based on Table 1, a fitting surface is plotted to find the optimal condition, shown in Figure 6.

The top point of the curve in Figure 6 with the highest success rate (98%) is the highest point in the whole figure. The condition where the percentage of cells storing information in cue-layer is 10% and the one in image matrix is also 10% is the optimal condition. Therefore, the test result is record which is in condition 10% density image matrix and 10% cue fulfillment. That means when the 10% of the brain space is used to store information, the effect of remembering association is best. That is why in human beings brain, 10% of the storing space is occupied.

Table 1. Successful rate.

Internal Matrix \ Cue Layer	10%	30%	50%	70%
1%	86%	68%	46%	30%
5%	96%	86%	74%	70%
10%	98%	82%	88%	76%
20%	90%	82%	90%	68%
30%	86%	84%	72%	74%
40%	74%	76%	70%	72%
50%	74%	76%	68%	68%
60%	66%	68%	70%	64%
70%	58%	70%	52%	70%
80%	60%	64%	58%	62%
90%	56%	44%	56%	56%

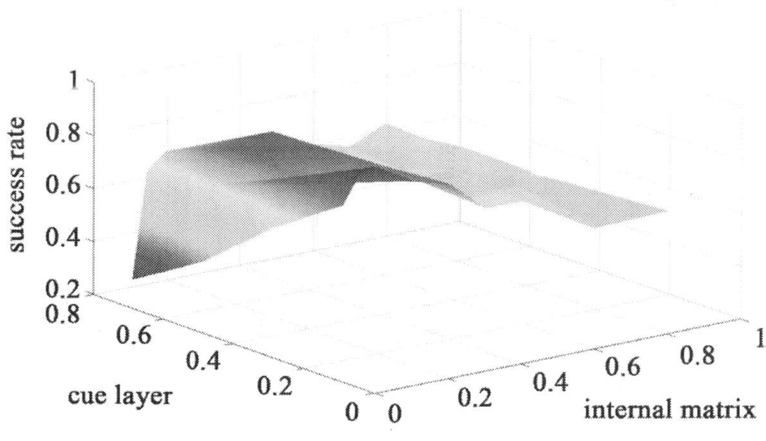

Figure 6. Fitting surface of successful rate.

However, other cells which not hold information are not useless and they are used to transfer excitation. In the course of human evolution, the people with a poor memory function are eliminated and the people with brain in the optimal condition are reserved.

Image what would happen if much more storage space is used or unbalanced storage space is used. If 70% of storage space is used both of character part and image part, the successful rate only 70%. That means if you told the patients that image represent an apple, in the next time he may call the other totally different image an apple [20] . Therefore the patients will be diagnosed for a psychopath whose memory is overloaded.

Moreover, in another condition where the two percentages of storage space of character part and image part are different, the function of brain is also influenced. For instance when the character capacity is 10% and the image part is 70%, the success rate is only 58%. That means if the man who has great talent in memorizing images but his capacity of character memorizing is not balanced with it of image memorizing, the man is also has a tendency to be insane. In other words, the genius who has gift on one aspect is nearly lunatic.

Generally, this model can be regarded as a mini-brain. The ability of study depends on the number of cells in use, which means that the length of the period for study decreases with the size of the matrix [21] . The biggest differences between this model and the real brain are the number of cells and the number of connections. Therefore, in this section, matrices with different dimensions and different numbers of connections are adjusted and then the results with different parameters are compared.

Firstly, the dimension of the matrix is expanded to 50 × 50, 80 × 80 and 100 × 100, respectively. Thus, correspondingly, the number of neural cells involved increases to 2500, 6400 and 10000, respectively. The condition with 10% density image matrix and 10% cue fulfillment (the optimal condition) are set to examine the effect of the size of the matrix on the successful rate. After that the case with dimension 50 × 50 has been repeated 50 times and a new successful rate 98% was obtained which is also comparatively high. The cases with dimension 80 80 and 100 × 100 were tested and the successful rates are 100% and 98%, respectively. Therefore, we can speculate that the successful rate may be independent from the size of matrix.

Secondly, in order to investigate whether the number of connections can affect the results of the experiment, we double the number of connections and observe, i.e., not only the up, below, left and right adjacent cell are connected to the central cell but also the up-right, up-left, bottom-left and bottom-right adjacent cell can be associated. Two pieces of different image information are given under the optimal condition. The relative cue- layers are different so the two pieces of image are successfully distinguished.

After that, the experiment has been repeated 50 times and the successful rate obtained is 100%. Therefore we can speculate that the number of connections may not influence the successful rate either.

According to the statistics, the dimension and the number of connections may have no effect on the successful rate. Therefore, we can conclude that the percentage of optimal occupied space in the brain is about 10%.

Finally, to simplify the computation and programming, the number of connections that a neuron cell could have with the adjacent cells is up to eight in this model. However, in fact, thousands of connections could be made by one single cell [21] .

Hence, further study could enlarge the matrix. Since in the model the matrix used to represent the processor has comparatively small dimensions, the processor is considered as a whole. When the matrix becomes larger and larger, partition of the processor could be taken into consideration.

In this model, only two states (excitatory and inhibitory) of a cell are considered. More values could be used to represent different states to increase the complexity of the situation. Moreover, the transmission is assumed to happen so quickly that forgetting can be neglected. In some other cases, forgetting may have significant effects on the memory process.

CONCLUSION

The model has been established to simulate the memory process of the human brain. Through conducting thousands of experiment by simulation, the statistical results indicate that the optimal percentage of occupation in the brain is 10%. However, the other cells which are not occupied are necessary for transmitting the excitation rather than unexploited. Furthermore, it can be speculated that the successful rate may be independent from the size of the matrix. And the dimension and number of connections may have no effect on the successful either.

ACKNOWLEDGEMENTS

This work was supported by grants from the National Natural Science Foundation of China (No.11204245), and the Natural Science Foundation of Jiangsu Province (No. BK2012637).

REFERENCES

1. Grossi, E. and Buscema, M. (2007) Introduction to Artificial Neural Networks. European Journal of Gastroenterology and Hepatology, 19, 1046-1054. http://dx.doi.org/10.1097/MEG.0b013e3282f198a0
2. Mastin, L. (2015) Long Term Memory. http://www.human-memory.net/types_long.html
3. Ma, C. and Zhang, N. (2015) Transforming Growth Factor Beta Signaling Is Constantly Shaping Memory T-Cell Population. Proceedings of the National Academy of Sciences of the United States of America, 112, 11013-11017. http://dx.doi.org/10.1073/pnas.1510119112
4. Vogels, T.P., Sprekeler, H., Zenke, F., Clopath, C. and Gerstner, W. (2011) Inhibitory Plasticity Balances Excitation and Inhibition in Sensory Pathways and Memory Networks. Science, 16, 1569-1573. http://dx.doi.org/10.1126/science.1211095
5. Hartmann, K., Bruehl, C., Golovko, T. and Draguhn, A. (2008) Fast Homeostatic Plasticity of Inhibition via Activity-Dependent Vesicularlling. PLoS ONE, 3, e2979. http://dx.doi.org/10.1371/journal.pone.0002979
6. Huxley, A.F. (1964) Excitation and Conduction in Nerve: Quantitative Analysis. Science, 145, 1154-1159. http://dx.doi.org/10.1126/science.145.3637.1154
7. Okun, M. and Lampl, I. (2008) Instantaneous Correlation of Excitation and Inhibition during Ongoing and Sensory-Evoked Activities. Nature Neuroscience, 11, 535-537. http://dx.doi.

org/10.1038/nn.2105

8. Luksys, G., et al. (2015) Computational Dissection of Human Episodic Memory Reveals Mental Process-Specific Genetic Profiles. Proceedings of the National Academy of Sciences of the United States of America, 112, 4939-4948. http://dx.doi.org/10.1073/pnas.1500860112
9. Kawato, M. and Gomi, H. (1992) The Cerebellum and VOR/OKR Learning Models. Trends in Neurosciences, 15, 445-453. http://dx.doi.org/10.1016/0166-2236(92)90008-v
10. Machens, C.K., Romo, R. and Brody, C.D. (2005) Flexible Control of Mutual Inhibition: Aneural Model of Two-Interval Discrimination. Science, 307, 1121-1124. http://dx.doi.org/10.1126/science.1104171
11. Nenert, R., Allendorfer, J.B. and Szaarski, J.P. (2014) A Model for Visual Memory Encoding. PLoS ONE, 9, e107761. http://dx.doi.org/10.1371/journal.pone.0107761
12. Flores, C.E., et al. (2015) Activity-Dependent Inhibitory Synapse Remodeling through Gephyrin Phosphorylation. Proceedings of the National Academy of Sciences of the United States of America, 112, E65-E72. http://dx.doi.org/10.1073/pnas.1411170112
13. Nielsona, D.M., Smitha, T.A., Sreekumara, V., Dennisa, S. and Sederberga, P.B. (2015) Human Hippocampus Represents Space and Time during Retrieval of Real-World Memories. Proceedings of the National Academy of Sciences of the United States of America, 112, 11078-11083.http://dx.doi.org/10.1073/pnas.1507104112
14. Mastin, L. (2010) The Human Memory. http://www.human-memory.net/types_short.html.
15. Nomura, H., et al. (2015) Memory Formation and Retrieval of Neuronal Silencing in the Auditory Cortex. Proceedings of the National Academy of Sciences of the United States of America, 112, 9740-9744. http://dx.doi.org/10.1073/pnas.1500869112
16. Markowitza, D.A., Curtisa, C.E. and Pesarana, B. (2015) Multiple Component Networks Support Working Memory in Prefrontal Cortex. Proceedings of the National Academy of Sciences of the United States of America, 112, 11084-11089. http://dx.doi.org/10.1073/pnas.1504172112

17. Wada, N., Funabiki, K. and Nakanishi, S. (2014) Role of Granule-Cell Transmission in Memory Trace of Cerebellum-Dependent Optokinetic Motor Learning. Proceedings of the National Academy of Sciences of the United States of America, 111, 5373-5378. http://dx.doi.org/10.1073/pnas.1402546111
18. Vreeswijk, C.V. and Sompolinsky, H. (1996) Chaos in Neuronal Networks with Balanced Excitatory and Inhibitory Activity. Science, 274, 1724-1726. http://dx.doi.org/10.1126/science.274.5293.1724
19. McTighe, S.M., Cowell, R.A., Winters, B.D., Bussey, T.J. and Saksida, L.M. (2010) Paradoxical False Memory for Objects after Brain Damage. Science, 330, 1408-1410. http://dx.doi.org/10.1126/science.1194780
20. Yger, P., Stimberg, M. and Brette, R. (2015) Fast Earning with Weak Synaptic Plasticity. The Journal of Neuroscience, 35, 13351-13362. http://dx.doi.org/10.1523/JNEUROSCI.0607-15.2015
21. Gerstner, W. and Naud, R. (2009) How Good Are Neuron Models. Science, 326, 379-380. http://dx.doi.org/10.1126/science.1181936

SECTION II:
TOPICS FROM NEUROSCIENCE

CHAPTER 5

DISCOVERING PATTERNS IN BRAIN SIGNALS USING DECISION TREES

Narusci S. Bastos, Diana F. Adamatti, and Cleo Z. Billa

Federal University of Rio Grande (FURG), Rio Grande, RS, Brazil

ABSTRACT

Even with emerging technologies, such as Brain-Computer Interfaces (BCI) systems, understanding how our brains work is a very difficult challenge. So we propose to use a data mining technique to help us in this task. As a case of study, we analyzed the brain's behaviour of blind people and sighted people in a spatial activity. There is a common belief that blind people compensate their lack of vision using the other senses. If an object is given to sighted people and we asked them to identify this object, probably the sense of vision will be the most determinant one. If the same experiment was repeated with blind people, they will have to use other senses to identify the object. In this work, we propose a methodology that uses decision trees (DT) to investigate the difference of how the brains of blind people and people with vision react against a spatial problem. We choose the DT algorithm because it can discover patterns in the brain signal, and its presentation is human interpretable. Our results show that using DT to analyze brain signals can help us to understand the brain's behaviour.

INTRODUCTION

Neuroscience is the scientific study of the nervous system. And with new technologies emerging, such as Brain-Computer Interface (BCI) systems, new discoveries can be made in this area. But analyzing and interpreting data collected from these BCI systems is not an easy job, so software tools have been developed in order to aid scientists in this area.

This work proposes a methodology that uses a data mining technique, called decision tree (DT), to discover patterns in brain signals and presents a model (a decision tree) that is easily interpretable. To show how our methodology works, we use a case scenario where we examine the brain signals of blind people and sighted people in spatial activity.

Blindness is a severe or total change of one or more elementary functions of vision; it affects the ability to perceive color, size, distance, shape, position, or movement in a given space [1]. The expression "visual impairment" refers to the spectrum ranging from blindness to low vision. There are two types of blind people: congenitally blind and acquired blind. The congenitally blind people have the cognitive system based on only 4 senses since birth, without any reference to visual elements, unlike the acquired blind ones, who have to adapt their cognitive system to their new condition [2].

The nervous system is responsible for the reception, storage, and release of information. It is a complex system consisting of various structures and specialized organs with different functions [3]. It can be divided into sensory system, which is responsible for collecting information about the organism and the environment; motor system, which organizes and executes actions; and the associative system. In this work, we focus our study on the sensory system, since it is known that individuals with visual impairments have their orientation capacity compromised [4].

BCI systems are tools (hardware and software) that allow a way of communication between the brain and the computer. They usually capture and process neural activities of the brain, and they do not require other stimuli, such as muscle movements [5]. There are different kinds of BCI; in this work, we used an EEG- (electroencephalogram-) based BCI system to collect brain signals. The EEG records brain electrical activity measured on the surface of the scalp and it is capable of capturing brain activity every millisecond due to a high temporal resolution. Most BCI systems provide only brainwaves and graphics (such as brain maps) tools to analyze the brain activity. In this work, we propose the use of data mining to analyze the brain's behaviour.

Data mining (DM) is the process of extracting or mining knowledge from a large volume of data and there are several algorithms that can be used to discover patterns in a dataset [6]. This work focuses on the use of decision trees for knowledge discovery in a dataset of brain signals of blind (visual impairment) and sighted people during an activity involving spatial abilities in order to discover whether there is a difference in their brain activity.

This work is divided into five sections. Section 2 presents a theoretical background, such as brain areas and their functions, visual impairment, BCI systems, and data mining. Section 3 provides the materials and methods we used. Section 4 presents the obtained results, and finally Section 5 presents conclusions and future works.

THEORETICAL BACKGROUND

Brain Areas and Their Functions

The brain is the main component of the nervous system. It is responsible for all mental operations such as concentration, thinking, learning, and motor control. These capabilities are implemented through neurons, which can currently be explained by neuroscience.

Human brain is divided into two hemispheres, right and left. Initially, there was a belief that there was one dominant hemisphere and the other was dominated. However, this concept has become outdated, and now there is a belief that there are actually two specialized hemispheres. Thus, each hemisphere is responsible for a set of functions that end up working together.

Anatomists usually divide the brain into major regions, called lobes, whose boundaries are not always accurate but transmit an initial idea of regional location. There are five lobes: four external and one internal, located in the lateral sulcus [7]. The four external lobes are the following: frontal lobe, which is located in the forehead; parietal lobe, which is located under the cranial bone with the same name; temporal lobe, which is associated with the tempora; and occipital lobe, which is located in the occipital cranial bone. The fifth lobe, the insula lobe, can only be seen when the lateral sulcus is opened [7, 8]. There are many other structures situated in the central nervous system (CNS), but in this work we investigate only the four visible lobes because the BCI system that we used does not have access to the insula lobe.

Each lobe has specialized functions: the occipital lobe is primarily concerned with the sense of vision; it is divided into multiple distinct visual areas, in which the biggest one is the primary visual cortex. The parietal lobe is partially dedicated to the sense of touch; it is responsible for body sensitivity functions and spatial recognition. The temporal lobe contains the primary auditory cortex; it processes audio data, specific aspects of vision, language understanding, and some aspects of memory. Finally, the frontal lobe is responsible for cognitive actions, memory, and movement [8, 9].

Visual Impairment

The visual impairment, in any degree, compromises a person's ability to orient and move in space with security and independence [10]. So, people with visual impairment or blindness compensate this visions lack of information using other senses: hearing, smell, touch, and taste [11].

Brain-Computer Interface Systems

BCI systems are a set of tools that enable communication between a brain and a computer. The main objective of BCI systems is to provide interaction between a user and an external device, such as computers, switches, or prostheses, using only brain signals. There are different ways to collect brain signals; one of them is to use the electroencephalography (EEG). The EEG is based on detecting brain electrical activity through electrodes applied to the scalp [12].

The signals that are captured by an EEG equipment are the potential differences between regions of the cortex. These electrical signs are generated due to the flow of ions between the different neurons of the brain. When a neuron is activated, it is polarized, generating an action potential that can be propagated to other neurons, provoking a flow of information [13].

The records acquired through the electrodes represent the intensity of brainwaves. They can vary between 0 μV and 200 μV, and they have frequency ranging from 0.3 Hz to 100 Hz. The resulting signal of an EEG shows peaks related to existence of electric activity, indicating a general spatial location of brain activity, because this signal is the sum of the activity of a large number of neurons communicating with each other [14].

Actichamp and Acticap

The Actichamp tool is developed by Brain Vision LLC. It is a modular amplification system that incorporates large components for electrophysiological analysis as EEG, event-related brain potentials (ERP), and BCI. It was used in conjunction with Acticap, which is a cap with 32 electrodes, and it is inserted into the scalp of a person. It has the channels of the international standard "10–20." The Acticap is connected to the Actichamp amplifier, to transmit the signals captured by the electrodes. Figure 1 shows how the electrodes are distributed throughout the cap.

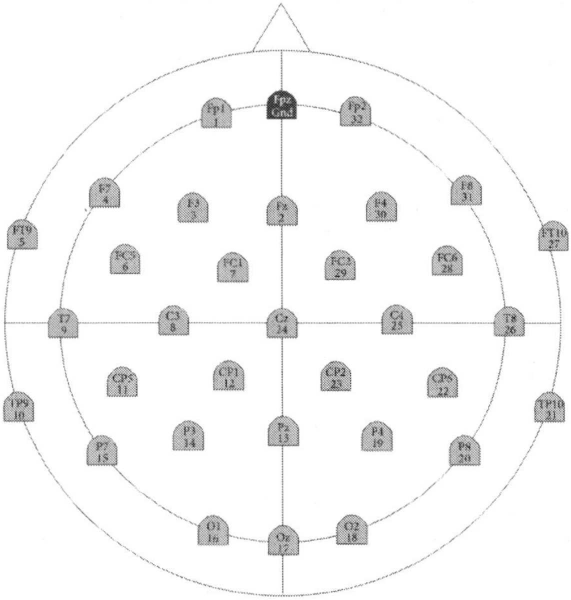

Figure 1: Location of the electrodes in Acticap.

The locations for each electrode are calculated to be in the intersection of the lines between standard cranium landmarks (see Figure 1). The name of each electrode indicates the region of the brain: FP indicates the prefrontal lobe; F, frontal lobe; T, temporal lobe; C, the central groove; P, parietal lobe; and O, occipital lobe. The number or the second letter identifies the hemispheric location: Z is the zero line in the center of the head; even numbers represent the right hemisphere; odd numbers represent the left hemisphere. The numbers are displayed in ascending order with increasing distance from the center [15, 16].

Table 1 shows the brain areas, the channels that constitute each area, and the abilities of each region.

Table 1: Brain region, electrodes, and proprietary functions.

Brain region	Electrode	Proprietary functions
Frontal lobe	Fp1, Fp2, Fz, F7, F3, Fz, F4, F8, FC5, FC1, FC2, FC6, FT9, FT10.	Executive functions (management of cognitive/emotional resources on a given task)
Temporal lobe	T7, TP9, T8, TP10.	Perception of biological motion
Parietal lobe	P7, P3, Pz, P4, P8.	Somatosensory perception, spatial representations, and tactile perceptions
Occipital lobe	O1, Oz, O2.	View images (including during a dialogue)

OpenVibe Software

OpenVibe is a software platform dedicated to designing, testing, and using Brain-Computer Interfaces. The configuration for use with Actichamp is predefined; the software communicates automatically with the signal capture tool. OpenVibe presents a very simple interface, where the user can set through an algorithm (automata) features that meet the needs of the task.

Data Mining

Data mining (DM) is the process of extracting or mining knowledge from a large volume of data. DM involves the study of tasks and techniques, where tasks are a specific class of problems and techniques are the groups of solutions to solve them [6].

Alencar et al. [17] point out that one of the most accepted definitions of data mining by researchers in the field is the one given by Fayyad et al. [18], which states the following: “database knowledge extraction is the process of identifying valid, new, potentially useful and understandable patterns embedded in the data.”

Data mining is one step in a broad process known as Knowledge Discovery in Database (KDD). KDD is the process of finding knowledge in data. In this context, DM is the step of obtaining the information [6].

Descriptive tasks are focused on discovering patterns that describe data in a way that human being can understand. The main descriptive tasks are association rules and clustering. Predictive tasks search for patterns to infer new information about the existing data or to predict the behaviour of new data. The main predictive tasks are classification and regression [6, 19].

The difference between predictive and descriptive methods consists in the fact that descriptive methods do not require a precategorization of records; that is, it is not necessary target in an instance; in predictive methods, the dataset has a predefined target variable and records are categorized in relation to it.

RELATED WORK

In the last years, a lot of work has been developed using data mining algorithms and, in some of them, these algorithms are used to perform classification tasks in brain signals.

Ishfaque et al. [20] report an experiment with several different classifiers in order to identify if a subject is moving his/her right hand forward or backward or his/her left hand forward or backward, thereby establishing a classification problem of 4 classes. In their experiment, they collect brain signals, using EEG-based BCI, of blindfolded subjects doing random movements of their right and left hands.

Ishfaque et al. [20] collected brain activities from 19 electrodes. The data was processed in the time domain, so their dimensionality was reduced from 19 columns to 5 columns using Principal Component Analysis (PCA). The amplitude of all four classes differentiated in time made the remaining data clearly separable.

After the data was collected, Ishfaque et al. [20] tested different types of classifiers in order to analyze the performance of each one. They use the following classifiers: Artificial Neural Networks (ANN), Linear Discriminant Analysis (LDA), and DT. To evaluate the results, they use a confusion matrix and accuracy percentage (AP). Both measures are used to check the accuracy of the classifiers.

According to the authors in [20], LDA, as expected, divided data linearly and because of that it did not present good classification ratings. On the other hand, ANN and DT reached good classification ratings, with ANN having better results. ANN accuracy percentage was 81.6%, DT accuracy percentage was 75.6%, and LDA accuracy percentage was 24.0%, as we can see in Table 2.

Table 2: Classifiers accuracy percentage [20].

Classifier	PA
ANN	81.6%
LDA	24.0%
DT	75.6%

Another work that investigates machine learning algorithms in order to classify brain signals is the one of Wang et al. [21]. They made an experiment where subjects had identified left or right arrows and pressed the corresponding key on the keyboard. They collected two datasets using an EEG-based BCI system.

In the work of Wang et al. [21], they tested the following classification algorithms: LDA, Quadratic Discriminant Analysis (QDA), Kernel Fisher Discriminant (KFD) analysis, Support Vector Machines (SVM), Multilayer Perceptron (MLP), Learning Vector Quantization (LVQ), ANN, K-Nearest Neighbors (KNN), and DT. The unit of measurement for evaluation was the accuracy.

Wang et al. [21] conclude that Gaussian SVM and KNN reached good performance ratings in both datasets, while LVQ, QDA, KDF, and MLP reached the lowest ratings. Linear SVM and LDA presented similar performance. Wang et al. [21] point out that KNN is not commonly used to classify brain signals, but with the appropriate resources extraction and reducing the vector's dimension, KNN can reach good performance ratings. Table 3 shows the reached accuracy percentage of each classifier in the experiments of Wang et al. [21].

Table 3: Percentage of the accuracy of the classifiers in different datasets [21].

Classifier	Dataset I PA	Dataset II PA
LDA	82.86%	84%
QDA	78.57%	79%
KFD	80.71%	81%
Linear SVM	82.86%	82%
Gaussian SVM	84.29%	84%
MLP	80.71%	81%
LVQ	77.86%	80%
K-NN	84.29%	83%
DT	82.14%	86%

The works of Ishfaque et al. [20] and Wang et al. [21] compare classification algorithms to evaluate which one has better results to classify brain signals in a specific domain. They propose the use of different kinds of classifiers and measure their accuracy. In this work, we propose to use a classification algorithm, specifically DT, to discover patterns of reasoning in the brain. In other words, we do not propose to use DT to classify if the subject is performing a specific action; we are interested in discovering how the brain behaves when a specific task is proposed, and to investigate this brain's behaviour, we used a DT algorithm to discover patterns in the brain signals.

METHODOLOGY

Figure 2 is the flowchart of our methodology. Each box is described in detail in the following subsections.

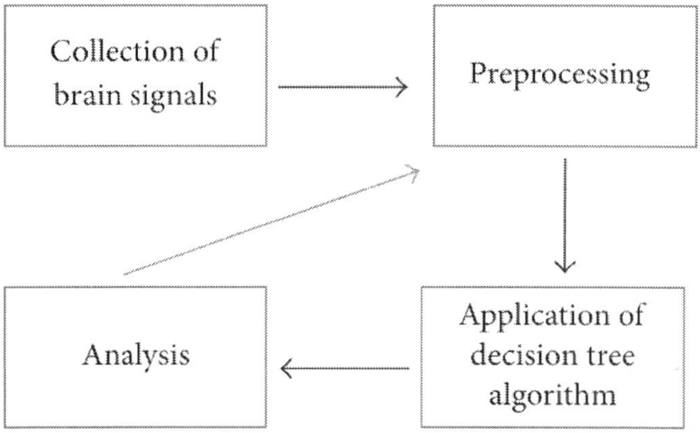

Figure 2: Methodology proposed in this work.

Collection of Brain Signals

We acquired the brain signals of 4 female individuals: 2 not blind and 2 blind people. The task we gave to them was to identify different 3D solid geometric shapes, in order to stimulate their spacial abilities. In our protocol, we used three objects: ball, cube, and parallelogram. All tests were performed with the approval of the Research Ethics Committee at the Health Area in Brazil, CCAAE: 344172114.3.0000.5324.

Detailed data collection is as follows:(1)The data was collected in a private room, with only the subject and the researchers.(2)The tools used to collect the brain signals were Actichamp and Acticap.(3)Calibration of the equipment: the electrodes have to be stimulated until they showed enough impedance to make it possible to start the collection.(4)The electrodes were connected to an audio recorder and the OpenVibe software was used for the acquisition and monitoring of brain signals.(5)The eyes of the subject were blindfolded.(6)Object one was given to the subject.(7)The subject handled the object and verbalized the name of the object.(8)Steps (6) and (7) were repeated with objects two and three.

Figure 3 shows the OpenVibe's automata used for brain signal acquisition and visual monitoring of them. The automata that follow the left side ending in "signal" display are performed, since the algorithms do not interfere with the signal acquisition. For signal acquisition, only the "acquisition client" and "GDF file writer" are used.

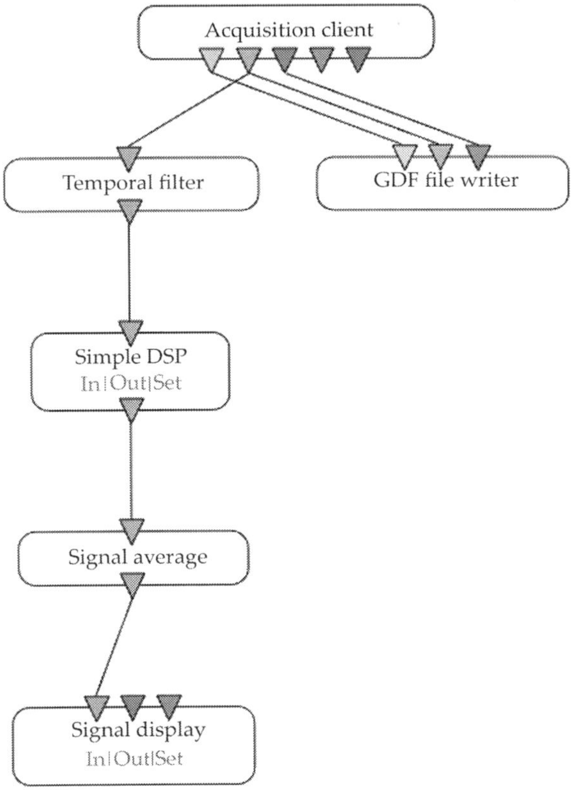

Figure 3: Automata used for the acquisition of brain signals.

The "acquisition client" waits for data from the EEG and it distributes the signals to the scenario. The algorithm opens a socket to read the experiment information, sign, stimulus, and channels location data sent across the network.

The "GDF file writer" is a function that writes to disk a specific current output in standard file format GDF (Graph Exploration System). This "box" does not allow changes (the user can just inform the file name to be saved).

Preprocessing

The main steps of data preprocessing are(1)conversion from GDF file to CSV file(2)balancing data(3)normalization(4)grouping.

Conversion from GDF File to CSV File

For the conversion from GDF to CSV, it was necessary to create a scenario in OpenVibe software also containing the filters required for analysis of brain signals (Figure 4). This transformation was necessary because the CSV files can be read directly by the Weka data mining tool.(1)"GDF file reader" has the following configuration:(a)Samples per buffer: 32(b) Subtracting physical minimum: false(2)"Temporal filter" is used to filter the input signal (Figure 5). The Butterworth filter is designed to introduce a flatter frequency response in the passband. The frequency ranges from 3.5 to 30 Hz, getting the theta, alpha, and beta waves.(3)"DSP filter" was set to to remove negative signs.(4)"Signal average" was used to calculate the mean of each input sample and outputs a resulting signal.(5)"CSV file writer" was used to record the filtered data in a CSV file.(6)"Signal display" was used for monitoring the data during format conversion.

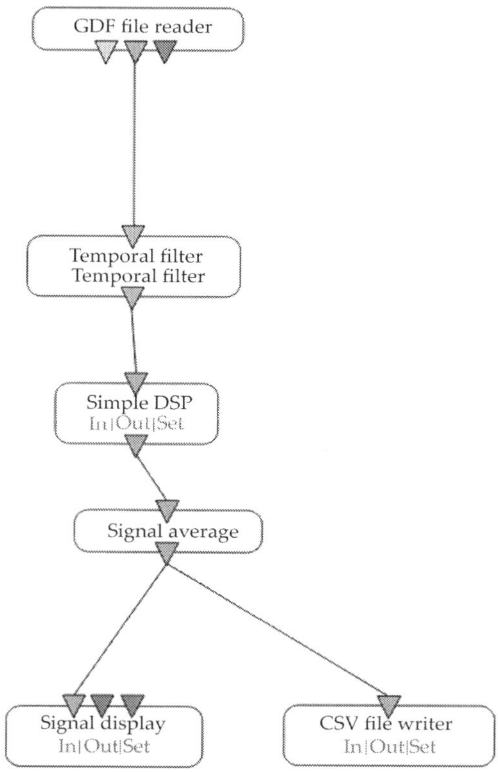

Figure 4: Automata to convert from GDF file to CSV file.

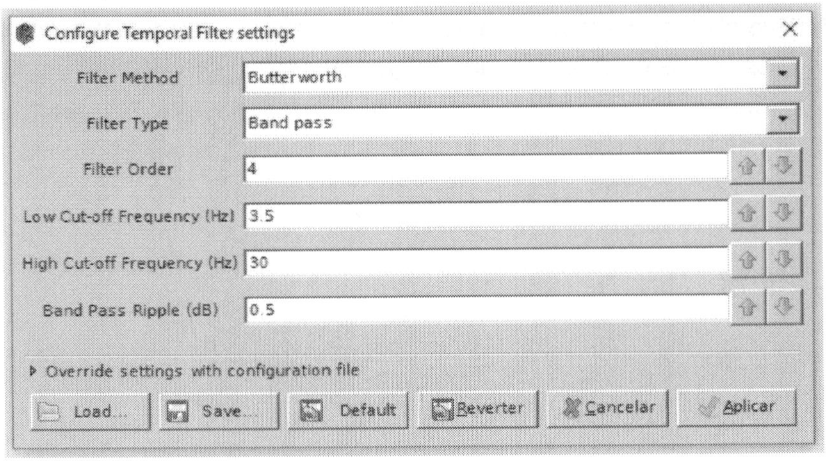

Figure 5: "Temporal filter" settings.

Balance

As a decision tree algorithm can be influenced by unbalanced data, we had to balance the data. Using the Weka software, we use a random filter to get 40 instances of each class. The number 40 was chosen because it was the number of instances of the class with less instances.

Normalization

The normalization step was done in Weka software, applying the filter "normalized," which transforms the values of the instances on a scale of 0 to 1.

Grouping

In order to identify temporal patterns, the data were grouped in a period of 1 second (10 instances). For each channel, the highest value was kept. So, at the end, each class had 4 instances, creating a new table with 16 instances.

Application of Decision Tree Algorithm

In this study, we have used the J48 decision tree algorithm, which is a classification algorithm. The J48 tree decision (also called C4.5) is an algorithm that uses the method of divide and conquer to increase the predictive ability of decision trees. In this way, it always uses the best step assessed locally, without worrying if this step will produce the best solution, takes a problem, and divides it into several subproblems, creating subtrees between the root and the leaves. We used the Weka software to execute the J48 algorithm.

The Weka (Waikato Environment for Knowledge Analysis) is a collection of machine learning algorithms for data mining tasks. It was developed by the Department of Computer Science at the University of Waikato, New Zealand [22]. These algorithms can be applied directly or used by Java programs. Weka contains algorithms for preprocessing, classification, regression, clustering, and association rules [19].

Analysis

In a decision tree, each leaf node receives a class label. The nonterminal nodes, including the root node and other internal nodes, contain attributes test conditions to separate records that have different characteristics [23]. Figures 6 and 7 show an example of trees generated by the J48 algorithm. From the generated tree, we can extract some rules as follows:(1)Time ≤ 14: cube.(2)Time > 14, time ≤ 30, and CP1 ≤ 0.072: interval.(3)Time > 14, time ≤ 30, and CP1 > 0.072: ball.(4)Time > 14 and time > 30: parallelogram.

```
Time ≤ 14: cube (4.0)
Time > 14
|    Time ≤ 30
|    |    Cp1 ≤ 0.072: interval (4.0)
|    |    Cp1 > 0.072: ball (4.0)
|    Time > 30: parallelogram (4.0)
Number of leaves: 4
Size of the tree: 7
```

Figure 6: Algorithmic representation of the decision tree.

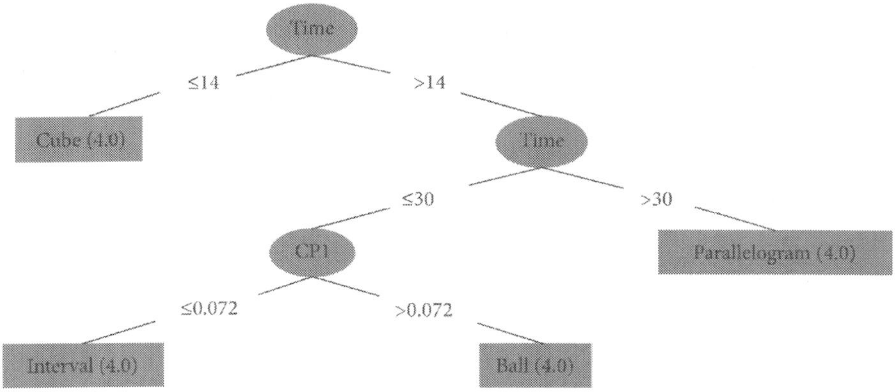

Figure 7: Graphical representation of the decision tree.

RESULTS

During our research, we performed several experiments using the J48 algorithm with different settings. Firstly, the full set of examples was tested. In second place, some electrodes that we do not consider relevant were cut. Thus, instead of 32 channels (all), only 12 channels were used.

Beyond the above-mentioned settings, we also executed tests by varying the minimum number of instances per leaf (minNumObj in Weka), which is a parameter of the J48 algorithm. We tested the values 1%, 5%, and 10% of the total number of instances. The other parameters were set with the default values of Weka, as shown in Figure 8.

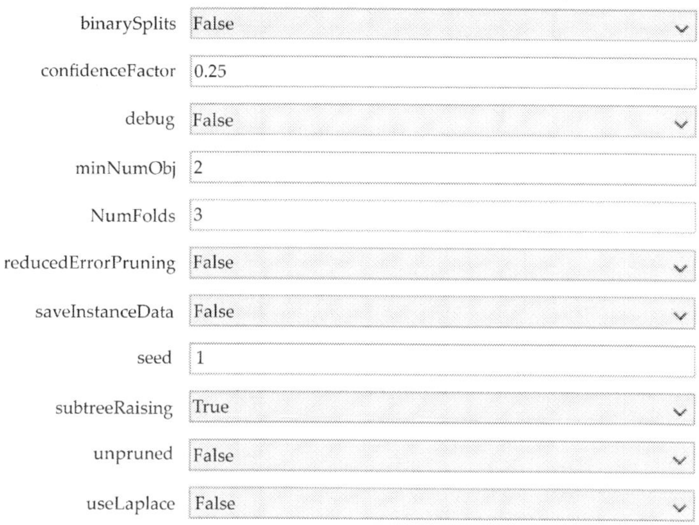

Figure 8: J48 parameters.

In most of these tests, we get good classification ratings, but the generated trees were very large and difficult to analyze. As our goal is not to classify new instances but to use decision trees to discover which areas of the brain had more significant activities, a tree with too many branches is not easy to analyze. We suppose that these trees were getting very large, probably, due to overfitting, since each tree leaf classified very few instances.

To avoid this overfitting, we grouped a subset of 10 consecutive instances. Each attribute received the highest value of its set (as mentioned in Section 4.2). So, in these tests, each instance contains the peak of each electrode in a larger period of time, representing in one instance a set with the highest values for each electrode during period of 1 second (approximately). With this configuration, we apply the DT algorithm to each subject's dataset. We executed them with the minimum number of objects set to 1%, because we had a small amount of instances of each class. The generated decision trees are shown in Figures 9, 10, 11, and 12 and they classify 100% correctly all the instances.

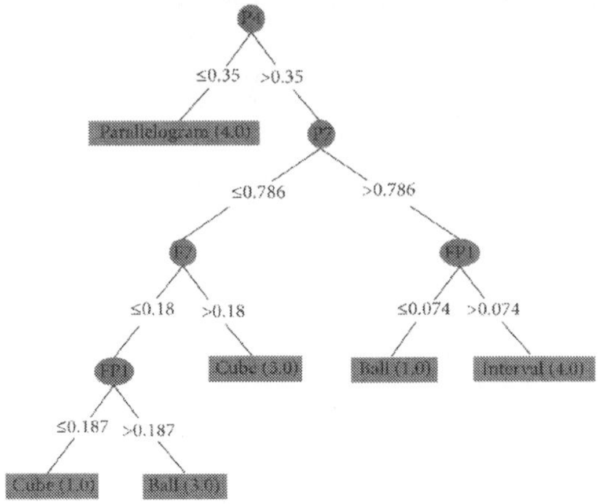

(a)

P4 ≤ 0.35: parallelogram (4.0)
P4 > 0.35
| P7 ≤ 0.786
| | F7 ≤ 0.18
| | | FP1 ≤ 0.187: cube (1.0)
| | | FP1 > 0.187: ball (3.0)
| | F7 > 0.18: cube (3.0)
| P7 > 0.786
| | FP1 ≤ 0.074: ball (1.0)
| | FP1 > 0.074: interval (4.0)

(b)

Figure 9: Blind person I: (a) DT's graphical representation and (b) DT's algorithmic representation.

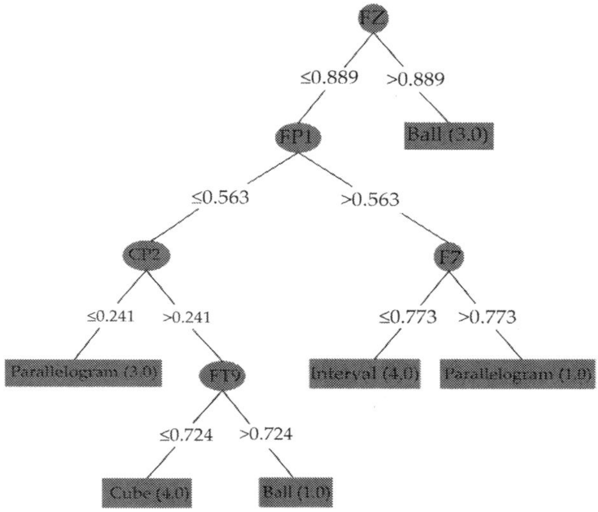

(a)

```
FZ ≤ 0.889
|   FP1 ≤ 0.563
|   |   CP2 ≤ 0.241: parallelogram (3.0)
|   |   CP2 > 0.241
|   |   |   FT9 ≤ 0.724: cube (4.0)
|   |   |   FT9 > 0.724: ball (1.0)
|   FP1 > 0.563
|   |   F7 ≤ 0.773: interval (4.0)
|   |   F7 > 0.773: parallelogram (1.0)
FZ > 0.889: ball (3.0)
```

(b)

Figure 10: Blind person II: (a) DT's graphical representation and (b) DT's algorithmic representation.

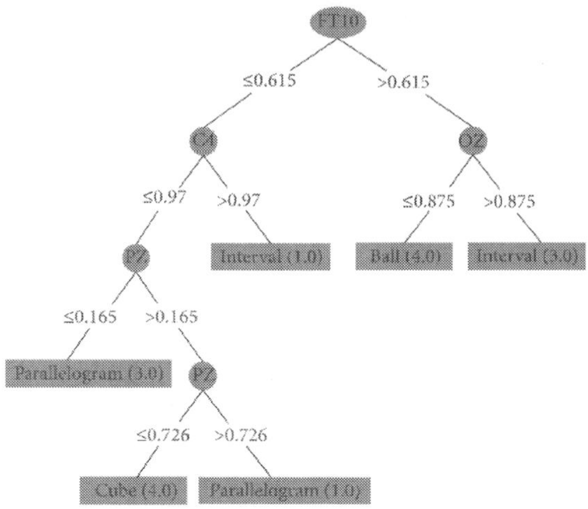

(a)

```
FT10 ≤ 0.615
|    C4 ≤ 0.97
|    |  PZ ≤ 0.165: parallelogram (3.0)
|    |  PZ > 0.165
|    |  |  PZ ≤ 0.726: cube (4.0)
|    |  |  PZ > 0.726: parallelogram (1.0)
|    C4 > 0.97: interval (1.0)
FT10 > 0.615
|    OZ ≤ 0.875: ball (4.0)
|    OZ > 0.875: interval (3.0)
```

(b)

Figure 11: Sighted person I: (a) DT's graphical representation and (b) DT's algorithmic representation.

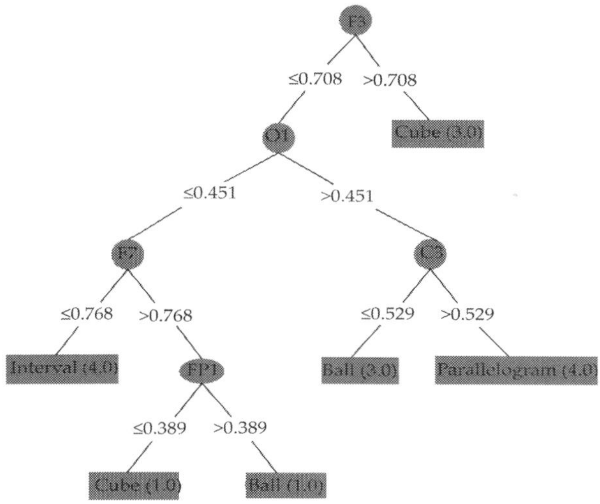

(b)

```
F3 ≤ 0.708
|   O1 ≤ 0.451
|   |   F7 ≤ 0.768: interval (4.0)
|   |   F7 > 0.768
|   |   |   FP1 ≤ 0.389: cube (1.0)
|   |   |   FP1 > 0.389: ball (1.0)
|   O1 > 0.451
|   |   C3 ≤ 0.529: ball (3.0)
|   |   C3 > 0.529: parallelogram (4.0)
F3 > 0.708: cube (3.0)
```

(b)

Figure 12: Sighted person II: (a) DT's graphical representation and (b) DT's algorithmic representation.

Figures 9, 10, 11, and 12 present the decision trees generated by the J48 algorithm, where (a) and (b) are two different ways to represent the same decision tree: (a) is a graphical representation of a tree and (b) is the algorithmic representation. Figures 9 and 10 represent the decision trees of two blind people. And Figures 11and 12 represent the trees of sighted people.

In the trees of visual impairment individuals (congenital blindness) (Figures 9 and 10), the trees show that the channels presenting the most significant activities were at the parietal lobes (Figure 9: P4, P7; Figure 10: CP2) and at the front lobe (Figure 9: F7, Fp1; Figure 10: Fz, Fp1, FT9, and F7). As we mentioned before (Table 1), the parietal lobe is responsible for coordinating actions that are sensitive to skin, as the sense of touch. The frontal lobe coordinates motor activities, thinking, and speech.

In the trees of sighted individuals (blindfolded), according to Figures 11 and 12, the channels that presented significant activity correspond to the frontal lobe (Figure 11: FT10; Figure 12: F3, F7, and Fp1), parietal lobes (Figure 11: Pz), occipital lobes (Figure 11: Oz; Figure 12: O1), and central lobe (Figure 11: C4; Figure 12: C3). We can notice that, in all tests, blind and sighted individuals showed high activity in the frontal and parietal lobes, where the first is responsible for the organization of the thoughts and the second is responsible for the sense of touch. However, sighted people showed significant activity in the occipital lobe, which is responsible for the sense of vision. In blind people, this area was not activated, because the J48 algorithm considered the values of the occipital lobe's electrodes insignificant for the generated model.

CONCLUSION

This work analyzed four individuals, including two not blind people and two people with congenital visual impairment, in order to analyze brain activity during a spatial activity. For this analysis, we collect the brain signals with Actichamp tool and process these data with the Weka software to data mining. We choose the J48 data mining classification technique because it generates decision trees that are easy to analyze.

Based on the resulting decision trees, we can observe that sighted people had significant activity in the occipital lobe, which is responsible for the sense of vision, even when they are blindfolded. We suppose that this happened because they accessed the visual memory to aid them to identify the objects. However, blind people showed no significant activity in the occipital lobe in the model created by the J48 algorithm. Therefore, our experience suggests that the brains of blind people and people with normal vision have different ways of carrying out spatial activities, even if they are placed in the same situation (people with vision were blindfolded).

As future works, we intend to expand the study with a larger number of people. We also intend to apply the methodology in other neuroscience studies, since DT can be used to discover patterns and understand brain signals in many different kinds of problems.

COMPETING INTERESTS

The authors declare that they have no competing interests.

REFERENCES

1. E. D. de Sá, I. M. de Campos, and M. B. C. Silva, Atendimento Educacional Especializado: Deficiência Visual, MEC, SEESP, 2007 (Portuguese).
2. V. Kastrup, "A Invention on fingertips: attention reversion in visually impaired people," Psicologia em Revista, vol. 13, no. 1, pp. 69–90, 2007.
3. C. Eyzaguirre, Physiology of the Nervous System. Year Book Medical, First Thus Edition, 1st edition, 1969.
4. M. Gil, Cadernos da TV Escola—Deficiência Visual, MEC-Secretaria de Educação a Distância, Brasília, Brasil, 2000 (Portuguese).
5. S. Machado, M. Cunha, B. Velasques et al., "Brain-computer interface: new prospects for rehabilitation," Revista Neurociências, vol. 17, no. 4, pp. 329–335, 2009 (Portuguese).
6. R. B. Pereira, A. Plastino, B. Zadrony, L. H. C. Merschmann, and A. A. Freitas, "Lazy attribute selection: choosing attributes at classification time," Intelligent Data Analysis, vol. 15, no. 5, pp. 715–732, 2011. ·
7. M. R. Lourenço, P. P. Garcez, R. Lent, and D. Uziel, "Temporal and spatial regulation of interneuron distribution in the developing cerebral cortex—an in vitro study," Neuroscience, vol. 201, pp. 357–365, 2012. View at Scopus

8. M. S. Gazzaniga, T. F. Heatherton, and M. A. V. Veronese, Psychological Science: Mind, Brain and Behavior, edited by W. W. Norton, 2006.
9. R. Lent and F. Tovar-Moll, "How can development and plasticity contribute to understanding evolution of the human brain?" Frontiers in Human Neuroscience, vol. 9, article 208, 2015. View at Scopus
10. Secretária de Educação Especial, Subsídios para a Organização e Funcionamento de serviços de educação especial, Ministério da Educação e do Desporto, Secretária de Educação Especial, Brasília, Brazil, 1995 (Portugueses).
11. S. D. Silveira Nunes and J. F. B. Lomonaco, "Desenvolvimento de conceitos em cegos congênitos: caminhos de aquisição do conhecimento," Psicologia Escolar e Educacional, vol. 12, no. 1, pp. 119–138, 2008 (Portuguese).
12. D. J. McFarland and J. R. Wolpaw, "Brain-computer interfaces for communication and control," Communications of the ACM, vol. 54, no. 5, pp. 60–66, 2011. View at Scopus
13. M. Kugler, I. N. Mukai, H. S. Lopes, and V. Pilla Junior, "Recognition of EEG patterns related to evoked potentials by intermittent photo-stimulation," in Proceedings of the International Conference on Mathematics and Engineering Techniques in Medicine and Biological Sciences, pp. 344–350, CSREA Press, Las Vegas, Nev, USA, 2001.
14. I. Oliveira, L. Carrico, N. Guimaraes, T. Chambel, and C. Teixeira, Interfaces Computador-Cérebro: Extracção e Processamento de Características de Electroencefalogramas, Department of Informatics, University of Lisbon, 2008 (Portuguese).
15. A. Vallabhaneni, T. Wang, and B. He, "Brain-computer interface," in Neural Engineering, B. He, Ed., pp. 85–121, Kluwer Academic Publishers, New York, NY, USA, 2005.
16. American Clinical Neurophysiology Society, "Guideline 5: guidelines for standard electrode position nomenclature," American Journal of Electroneurodiagnostic Technology, vol. 46, no. 3, pp. 222–225, 2006.
17. A. B. Alencar, K. Börner, F. V. Paulovich, and M. C. F. de Oliveira, "Time-aware visualization of document collections," in Proceedings of the 27th Annual ACM Symposium on Applied Computing (SAC

'12), pp. 997–1004, ACM, Trento, Italy, March 2012. View at Scopus

18. U. Fayyad, G. Piatetsky-Shapiro, and P. Smyth, "From data mining to knowledge discovery in databases," AI Magazine, vol. 17, no. 3, pp. 37–53, 1996. View at Scopus
19. U. Fayyad, Advances in Knowledge Discovery and Data Mining, AAAI Press, Menlo Park, Calif, USA, 1st edition, 1996.
20. A. Ishfaque, A. J. Awan, N. Rashid, and J. Iqbal, "Evaluation of ANN, LDA and decision trees for EEG based brain computer interface," in Proceedings of the IEEE 9th International Conference on Emerging Technologies (ICET '13), pp. 1–6, IEEE, Islamabad, Pakistan, December 2013. View at Scopus
21. B. Wang, C. M. Wong, F. Wan, P. U. Mak, P. I. Mak, and M. I. Vai, "Comparison of different classification methods for EEG-based brain computer interfaces: a case study," in Proceedings of the IEEE International Conference on Information and Automation (ICIA '09), pp. 1416–1421, Zhuhai, China, June 2009. View at Scopus
22. J. Ferreira, Data mining em banco de dados de eletrocardiograma [Tese de Douturado Instituto Dante Pazzanese de Cardiologia], Universidade de São Paulo, 2014 (Portuguese).
23. P. N. Tan, M. Steinbach, and V. Kumar, Introduction to Data Mining, vol. 1, Pearson Addison Wesley, Boston, Mass, USA, 2006.

CHAPTER
6

THE BRAINARIUM: AN INTERACTIVE IMMERSIVE TOOL FOR BRAIN EDUCATION, ART, AND NEUROTHERAPY

Romain Grandchamp[1, 2] and Arnaud Delorme[3, 4, 5]

[1]Laboratoire de Psychologie et NeuroCognition, Université de Grenoble, Grenoble, BSHM, 1251 av Centrale CS40700, 38058 Grenoble Cedex 9, France

[2]CNRS, UMR 5105, Grenoble, France

[3]Centre de Recherche Cerveau et Cognition (CerCo), Université Paul Sabatier, Pavillon Baudot, Hopital Purpan, BP 25202, 31052 Toulouse Cedex 3, France

[4]CNRS, UMR 5549, Toulouse, France

[5]Swartz Center for Computational Neuroscience, Institute of Neural Computation (INC), University of San Diego California, La Jolla, CA 92093-0559, USA

ABSTRACT

Recent theoretical and technological advances in neuroimaging techniques now allow brain electrical activity to be recorded using affordable and user-friendly equipment for nonscientist end-users. An increasing number of educators and artists have begun using electroencephalogram (EEG) to control multimedia and live artistic contents. In this paper, we introduce a new concept based on brain computer interface (BCI) technologies: the Brainarium. The Brainarium is a new pedagogical and artistic tool, which can deliver and illustrate scientific knowledge, as well as a new framework for scientific exploration. The Brainarium consists of a portable planetarium device that is being used as brain metaphor. This is done by projecting multimedia content on the planetarium dome and displaying EEG data recorded from a subject in real time using Brain Machine Interface (BMI) technologies. The system has been demonstrated through several performances involving an interaction between the subject controlling the BMI, a musician, and the audience during series of exhibitions and workshops in schools. We report here feedback from 134 participants who filled questionnaires to rate their experiences. Our results show improved subjective learning compared to conventional methods, improved entertainment value, improved absorption into the material being presented, and little discomfort.

INTRODUCTION

This century has been marked by the development of new brain imaging techniques, which have allowed us to better understand how our brain functions when we experience different mental states. The brain appears as a key integrative organ where a variety of inputs are simultaneously processed and combined: exteroceptive stimuli, that is, stimulations coming from the external world, proprioceptive inputs that provide body state information, or interoceptive inputs such as thoughts, emotions, and other inner experiences [1]. This processing is the result of lifelong learning, shaping, and adaptation of our neural system through our interaction with the world [2].

With the discovery of some of the core processes underlying brain electrical activity, we have found a new way of looking at the brain of living beings, obtaining insights on the functioning of their perceptual and inner spaces. When groups of several thousand neurons in the brain, sitting at close distances from each other and oriented in the same direction, are firing synchronously, their joint electrical activity adds up and generates an electrical field that is strong enough to be captured on the scalp. As a result of technological developments in electronics, signal processing, and computer science, we are now able to record different electrical brain rhythms with millisecond precision and process this activity in real time by placing electrodes on subjects' head. The technique of electroencephalography (EEG) is now widely used both in fundamental and in clinical research, as well as a diagnostic tool in clinical environment. In addition to basic research and clinical applications, EEG rhythms have been recently used to control computers in real time. In Brain Machine Interface (BMI) or Brain Computer Interface (BCI) [3–5] characteristic patterns of EEG activity during specific mental activity are mapped to a given computer command. Some BCI systems allow controlling a mechanical device, a graphical interface, or a video game using thoughts only. Subjects may voluntarily learn to retrain specific brain EEG patterns in order to correct pathological activity of the brain. This specific range of application is called neurofeedback or neurotherapy [6–8].

When EEG began to be recorded in the 1930s, researchers realized that several typical rhythms could be distinguished in the brain electrical activity recorded at the surface of the scalp. The first "brainwave" was identified by the father of electroencephalography, Berger [9]. It was denominated using the first Greek letter, alpha, and became the "alpha" rhythm, which is a brain rhythm that oscillates at about 10 cycles per second (10 Hertz). This rhythm is particularly active when a person is awake, resting with his/her eyes closed or while relaxing [10, 11]. Using alpha brainwaves to create or modulate sound and/or music has been pioneered by Lucier [12] as recently as 1965. Later in 1969, Kamiya showed that it was possible to voluntary control the alpha brain rhythm and modulate audio feedback in real time [13].

Following technical and theoretical progresses in neuroscience, computer science, and signal processing, EEG signals have recently been used in new ways [14, 15]. With the development of affordable and user-friendly EEG systems, the last few years have seen an increasing number of art projects using brain electrical activity as an input or way to produce or modulate artistic content such as computer graphics, animations, music, and choreography. Several performances have been created around the concept of music generation using brainwaves [16–19]. The Global Mind Project (http://www.globalmindproject.com/) is an example of such an artistic project. This system allowed for audio-video rendering of brain data, which, when combined with live interactive performance, has helped further develop new interactive artistic productions. According to Clarke, an Honorary Fellow in the Department of Culture and Communication at the University of Melbourne, "drawn together in a coalescence of self and technology, the artists connected to the EEG headsets are presented as both automata – self-operating machines – and intentional, self-activating beings, that have the ability to affect and be affected by the on-screen imagery generated" [20].

Another recent realization developed by a team of Rensselaer Polytechnic Institute students is Yehuda Duenyas' Infinity Simulator which involves control of a 3D automated rigging system using specific brainwave patterns [21]. This device led to the creation of the Ascent project (http://theascent.co/), a live-action, participatory theatrical experience that combines mind-control and levitation via an automated custom-built lifting platform system.

Our system uses similar ideas with the important addition of an immersive environment. This is the first time to our knowledge that real time EEG recordings are being displayed in a full dome immersive environmentallowing direct spatialization (spatial transposition) of brainwave data. One of the main originality and strengths of our system is also the brain metaphor regarding the shape of the device. Among developed applications, it allows projecting EEG topographic activity directly on the dome surface of a planetarium as if viewers were standing at the centre of the brain, looking up at electrical brain activity projected on the scalp.

The "Brainarium" (originally "Cerveaurium" in French) was initially designed to present neuroscience concepts in a fun, attractive, and interactive way for educational and entertainment purposes, by mixing art and science. In the method section of this paper, we will first present the general concept and architecture of the system in order to outline and illustrate its general functioning, describe a first performance that was designed for the Brainarium, and detail the specific implementation of the performance using open source software platforms. In the Result and Critical Reception section, we then mention the context in which our device has been used during exhibitions at museums and during "The

Brain's Awareness Week" and present data on audience's experience in the Brainarium. In Discussion, we finally introduce potential extension and further development of our system in the fields of education, entertainment, arts, and more specifically its possible benefits in clinical applications such as neurotherapy.

METHODS

Concept and General Design

Figures 1 and 2 summarize the architecture of the system and the different hardware it is comprised of. EEG signal is acquired on a person present in the dome. As shown in these figures, we used the Emotiv Epoc headset (Emotiv, Inc.), which includes 14 metal electrodes recording electrical brainwaves on the surface of the scalp at a frequency of 240 Hz (240 samples per second), but any EEG system compatible with BCI software can potentially be used. The signal is then transmitted, using a wireless connection, to a computer. This processing unit handles the signal processing part and calculates the control signals, which will be used to drive multimedia contents. Visual representations are finally projected onto a planetarium dome via a video projector equipped with a hemispherical lens (the system can be adapted to project on a hemispheric mirror which will reflect the image on the dome surface rather than directly project on the dome using a hemispheric lens). The system can be upgraded to multiprojector full dome systems but the main advantage of using a transportable inflatable dome and monoprojector hemispheric projection system is that it decreases the overall cost and allows an itinerant use. The computer display adapter should have two video outputs in order to allow simultaneous control of the different software on one screen and output to the video projector for the dome. In addition, a video splitter was used to send the video signal to a second screen so that the person driving the performance could see what was being projected.

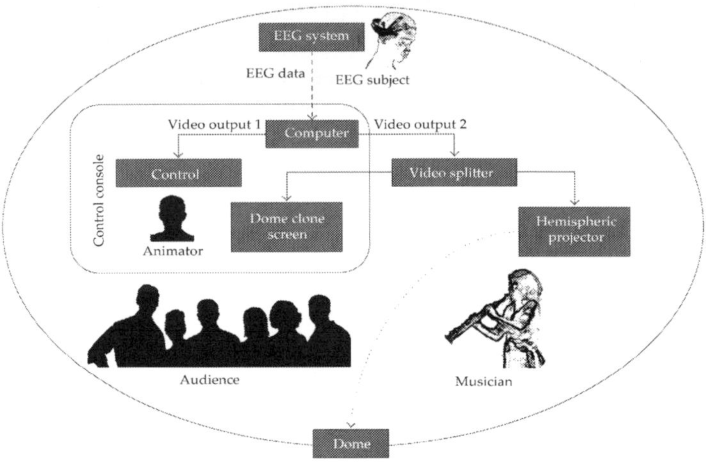

Figure 1: General principle of the Brainarium. EEG is recorded using the Emotiv headset and sent to a computer that computes brain rhythm activity in real time and projects it on the planetarium dome.

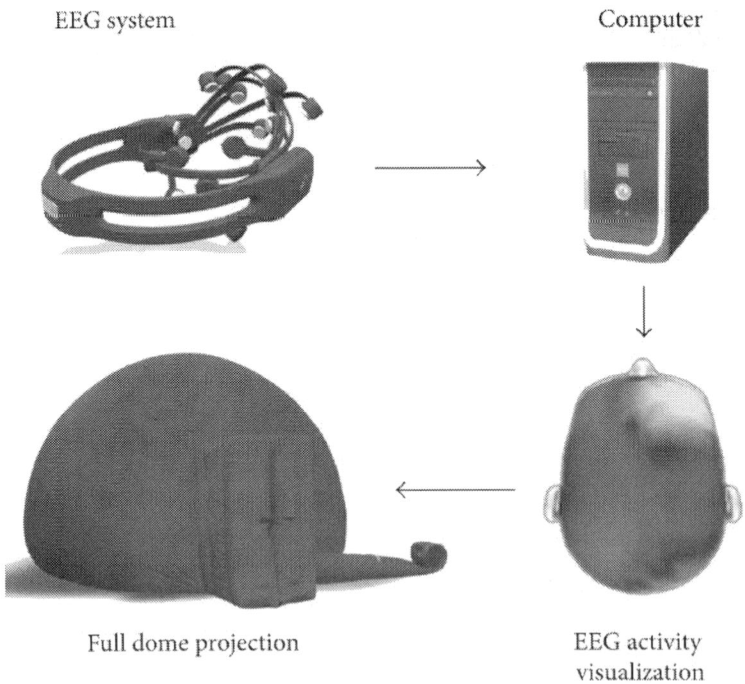

Figure 2: Flow chart of the different modules of the Brainarium.

If the system is used in the context of an art and science performance, brain electrical activity may be recorded from a member of the audience, an organizer of the projection, or an artist who participates in the event. Our device opens a wide range of possibilities among which we have integrated and used the following for a first performance:(i)Interaction with animation in computer graphics through electrical brain(s) activity(ies). (ii)Visualization of a brain rhythm (Alpha rhythm) associated to the suppression of visual input when the subject closes his eyes or relaxes. (iii)Real time presentation of topographies of brain electrical activity.(iv) Interactive presentation of brain's structures on a 3D brain model.

After presenting the general implementation of the system, the next sections will be devoted to description of each one of these applications.

General Implementation

Our set-up is based on combining a hemispheric projection system such as the one used in a planetarium, a hemispheric projection surface, and a brain computer interface system. Since every functional block of the system is modular, various solutions may be developed depending on budgetary constraints and available material. As we are writing this paper, the cost of building such a system could range from about US$5,000 to about a hundred thousand dollars when using research grade apparatus; the intermediate set-up we present here costs about US$40,000 although we also provide suggestions on how to build a similar system for a lesser amount.

For projection, we used a transportable planetarium system, which comprised a Digitarium® Delta Portable Digital Planetarium System [22] and a Digitalis™ Portable Dome [23] which has a diameter of 7 meters. However, both hemispheric projection systems and projection surfaces may be made at a lesser cost using custom made tools [24–26]. We implemented a low cost solution to replace the Digitarium Delta Portable Digital Planetarium System. This solution is composed of four parts: a full HD video projector (Acer H7531D), a condensator (Rodenstock TV Heligon 75 mm F/D = 1.1 can be replaced by a classical 50 mm with F/D = 1.4 combined with a +4 diopters lens as well), a 50 mm 45° mirror mount (Skywatcher), and a fisheye lens (Peleng 8 mm f3.5 fisheye lens). The dome we used was made of a thick fabric inflated by a powerful fan. This solution makes it more convenient to transport and set up the system compared to a rigid dome solution. However, this method has the drawback of having to leave the fan turned on in order to keep the dome inflated. Even if the sound of the fan is not covering the sounds played inside the dome, it still creates a distracting background noise.

The control console was composed of a classic personal computer equipped with a dual screen graphic card powerful enough to handle HD projection and two LCD monitors. One of the LCD monitors was used to control the demonstration. On the second graphical output, a video splitter was used to send the display signal to both a control LCD monitor and the video projector. For the EEG signal acquisition, the research edition package of Emotiv Epoc headset was used [27]. Emotiv Epoc is a wearable EEG "headset" composed of 14 gold-plated electrodes. In order to record electrical signals generated by the brain, each electrode is covered by a small felt-based pellet that acts as a bridge between the electrode and the scalp. These pellets have to be soaked in a saline solution, of water mixed with salt, which allows electrical conduction from the skin to the metal electrode through the pellet. The advantages of using this system are that it is relatively low in cost compared to clinical or research oriented devices. It is also wireless, fast, and easy to set up and provides some level of spatial resolution since it has 14 electrodes. However, clinical or research EEG systems with better signal quality can be used if available. Dry active electrodes would be the most adapted for such a system as they provide acceptable signal quality with a minimum preparation time but they are still expensive to date compared to the Emotiv Epoc solution.

The complete list of software used to run the system is depicted in Table 1. Except from the Emotiv software suite (the basic software package provided with the Epoc headset by Emotiv), the software used to do signal processing and visualization is all part of the open source community. For the fractal application, the software package "Mind Your OSC" was used to collect data from Emotiv Control Panel software and send it as an Open Sound Control (OSC) [28] stream to visualization software. The interactive fractal video was displayed using the vvvv software (https://vvvv.org/), a graphical programming environment for easy prototyping and development. The vvvv software application is designed to facilitate handling of large media environments with physical interfaces, real time motion graphics, audio, and video that can simultaneously interact with many users. The freely available OpenVibe software [29] was used for signal acquisition, signal processing, and visualization of the EEG data in the context of the EEG topography application. Finally, the 3D brain model application has been developed using Blender (https://www.blender.org/), a free open source 3D content creation suite, and rendered by its embedded real time full dome plugin [30]. We are making available all additional plugin and software developed for our application under an open source license [31].

Table 1: List of hardware and software applications required for setting up the Brainarium.

Software	Operating system	Function
Emotiv Control Center	MS Windows	Acquire EEG data and transmit it to the software "Mind Your OSC"
Mind Your OSC	MS Windows	Receive data from Emotiv Control Center and transmit them to vvvv using OSC protocol
vvvv	MS Windows	(i) Receive OSC data packets from "Mind Your OSC" (ii) Calibrate the system (iii) Compute the video speed (iv) Display hemispheric video
OpenVibe Acquisition Server	MS Windows or Linux	Acquire EEG data and transmit it to OpenVibe Designer
OpenVibe Designer	MS Windows or Linux	(i) Collect data from OpenVibe Acquisition Server (ii) Process EEG signal (extract alpha frequency band) (iii) Compute and display real time EEG topography
Blender	MS Windows or Linux	Display a 3D brain model in the Game Engine with a full dome display mode

Performance Design and Implementation

In this section we detail technical implementation of each application used for the different phases of the original performance designed for the Brainarium.

Brain-Controlled Animation of Fractals

This application is an example of live interaction. Figure 3 depicts the general architecture of the hardware and software for this application.

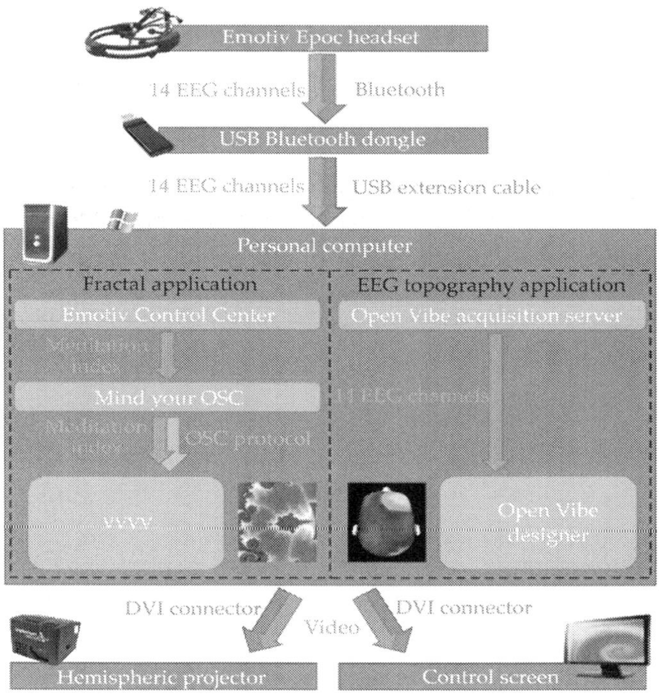

Figure 3: Set-up of the Brainarium for two different application examples: the fractal dynamical zoom application and the EEG topography application.

We first placed the EEG cap on the subject's head. Ideally, the subject's alpha brainwaves should be large compared to the overall electromagnetic noise. Since individual brains show different electrical rhythmic activities, some subjects can exhibit low amplitude alpha oscillations and this might make it more difficult to process the signal without the use of advanced artifact rejection techniques. Due to time limitation between sessions, we often asked a preselected person with known high amplitude alpha rhythm (i.e., easily observable on the signal trace) to be the subject.

After checking electrodes contact quality and signal quality, a calibration step lasting approximately two minutes is performed in order to evaluate some statistical features of the signal of interest's amplitude for the selected subject such as its mean and standard deviation. We used the index "Meditation" provided by the Emotiv Control Panel as the control signal. Since it has not been made public by Emotiv, we do not have the exact formula used to compute this index out of the raw EEG signal. However, it is known to be positively correlated with the alpha rhythm and relaxation. Emotiv indexes result from a statistical analysis based on a large normative database collected of many subjects and are therefore already normalized. However, a calibration procedure is still used in order to adapt the system to subject's specific statistics. In our case, we used the standard deviation and the mean value of the "Meditation" index over the calibration period as a reference value to tune the feedback set-up.

During the first minute of calibration we asked the subject to keep his eyes open and during the second minute we ask him to keep his eyes closed. Even if the subject has already performed the experiment, it is important to repeat the calibration step since EEG features widely vary throughout the day and from one day to another. "Meditation" values are calculated for both the eyes-closed and the eyes-open period and are used to calibrate the system to allow balanced behaviour of the visual feedback animation. Once calibration is performed, the session starts with the video feedback being projected on the dome and the audience enters the dome. In the meantime, a professional musician is improvising based on the visual display. This allows creating a complete interactive feedback loop between the subject wearing the EEG device and the musician (Figure 4(a)). The musician uses inspiration of what he sees on the dome to play music and adapt it. Furthermore, he can engage in an interplay with the wearer of the EEG and can try to induce changes in what is displayed.

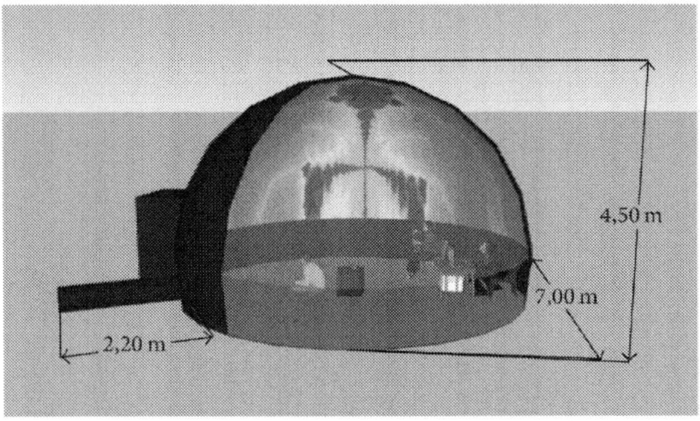

(a)

(b)

Figure 4: Brainarium represented as a 3D model with exact dimensions (the public is not shown on the rendering). (a) shows the projection of fractals and (b) shows the projection of subject's scalp topographies.

The "Meditation" measure controls the display projected on the dome. We used a zoom into Mandelbrot's ensemble fractal as visual feedback. More details about the video used are given in following paragraphs; we focus here on the interaction configuration. The speed and direction (forward or backward) of the zoom depends on the brainwaves of the subject wearing the EEG cap. The system was set up so that the animation was played forward, as if diving or moving forward into the fractal, when the current alpha wave amplitude generated by the subject was over its mean level. By contrast, when the current alpha wave amplitude was lower than its mean level, the animation was played backward, as if travelling away from the fractal. The speed of the animation was modulated by the difference between the current value of the alpha "Meditation" wave amplitude and its mean amplitude; that is, zooming becomes faster as the current value is further away from the mean. As a result, a "Meditation" value equal to the mean value would result in a static image.

The shapes projected on the dome are fractals. A 2D fractal is a mathematical expression, which may be represented as a 2D image. We choose to use fractals because, in addition to their aesthetic dimension, more and more research is showing that certain aspects of brain activity or even its own structure share some features with fractals [32–35]. Because fractals are based on mathematical expressions, there is no theoretical limit to the resolution of fractal images making it possible to zoom in on a small portion of the image and expanding it indefinitely. Another feature of fractal images is that their structure is preserved regardless of the "zoom." Finally, fractal images are self-similar representations. If the appropriate "zoom" is applied to a fractal image, the same image may be found again. An interesting feature resulting from the use of a fractal animation is that it produces an immersive tunnelling effect.

Fractal images presented in the Brainarium were made dynamic by zooming in or out in the fractal image. The animation used in the Brainarium was "a precalculated journey into the heart of the Mandelbrot fractal set" (http://www.hd-fractals.com/), which is named after Benoit Mandelbrot, the mathematician who studied and popularized it [36]. The video used in our demonstration features a 760 zoom in the Mandelbrot fractal set and it was produced by Teamfresh (http://www.hd-fractals.com/), an independent production company which specialized itself in rendering fractal animations. We used a commercially available High Definition version of the animation. The fractal video control application has been specifically implemented for this project using vvvv, a graphical programming environment for easy prototyping and development (https://vvvv.org/). We have made the vvvv patches developed for this application freely available [31].

EEG Real Time Topography Application

Specific software for this application can be seen in Figure 3. During the second part of the performance, participants observe EEG raw brainwaves, followed by their representation as topography or how

EEG brainwaves are distributed on the surfaces of the scalp. These EEG topographies may be likened to topographies used in elevation maps for hiking. Instead of representing the terrain elevation on the Earth surface, colors represent the strength of a specific brainwave at different locations on the head surface. In our case, we focused on brainwaves in a frequency band ranging from 8 to 12 Hz called the alpha band. Alpha brainwave amplitudes vary quickly in time and space and this dynamic may be rendered as animated colored maps on the dome. The topography is represented using either classic 2D spherical projection or an interactive 3D head model from OpenVibe software [29]. Using this set-up, participants may observe that when the subject closes his eyes, alpha wave amplitudes increase on the part of the dome that represents the back of the head. The part of the brain that is activated is called the occipital region, which is a brain area largely devoted to visual processing. When this region does not process visual information, that is, when the subject closes his eyes, alpha waves tend to increase in this brain area. Another way to increase alpha wave amplitude over the entire brain is to ask the subject to enter a deep relaxation state but this requires more training from the subject and this is more difficult to achieve in a single session: we have succeeded to perform the second part of this demonstration with only a few subjects. While the brain dynamic is shown on the dome, a musician is simultaneously playing his instrument, trying to help the subject to go into deeper relaxation states and simultaneously giving him auditory feedback about his relaxation state using his own interpretation of ongoing EEG patterns (Figure 4(b)). We have made available under an open source license the OpenVibe software scenario we developed to display alpha wave topography [31].

Neuroanatomy Using a 3D Interactive Brain Model

After the two interactive real time EEG sessions, the last part of our demonstration interactively showed different parts of the cortex in human brain volume. Despite the BCI being not involved in this part, we still want to describe it briefly to keep the description of the system's features complete. On the basis of gross topographical conventions, the cortex can be classified into four lobes: the temporal lobe, occipital lobe, parietal lobe, and frontal lobe. The system developed using Blender Game Engine (https://www.blender.org/) allows manipulating the 3D model in order to show the different lobes and introduce some of basic neuroanatomy concepts. We implemented rotation around different axis, zooming in and out for projection of these 3D models on the dome. The 3D models are rendered using the "Blender embedded full dome plugin" to compensate deformation due to the dome-specific projection lens and surface. We are making publicly available the Blender file we developed [31].

RESULTS AND CRITICAL RECEPTION

The innovative aspect of our project was to combine real time brain electrical activity visualization tools with an immersive full dome environment. Participants were seated inside the space enclosed by the projection dome, which induces a special atmosphere and feeling. In addition, scientific and artistic content interactive display exploited the analogy between the shape of the projection space and the near spherical shape of the brain (see Methods). What participants heard was not necessarily limited to what was being played inside the dome, as the material used for the projection surface was not soundproof. Nevertheless, acoustic properties of the dome were specific to its hemispheric shape, and this tended to enhance participants' experiences.

The Brainarium was inaugurated during "The Brain's Awareness Week," an event organized every year in all large European cities. For a week, series of exhibits are set up to present to the general public the latest advances in brain research. During "The Brain's Awareness Week" 2013, we performed more than 17 sessions demonstrating the Brainarium to more than 200 visitors. Following this encouraging start, our demo was also presented in Paris during the Cognitive Sciences Forum in the "Couvent des Cordeliers," at the Medical School of Paris, where it proved to be a very popular animation with more than 180 visitors in one day. Our project was also featured on the most popular newspaper of South-West France (6 million readers), "La Depeche," and also mentioned on local radio stations. It is now regularly requested for performances in more and more cities across France and Belgium, for workshops in primary and secondary schools, and for various national events such as the French National Science Week.

A questionnaire was filled in by participants after the performance to collect their feelings and how their experience in the Brainarium compares to traditional conferences and lectures they attended.

This questionnaire allowed us to collect demographic data about participants, on four closed questions with Likert scales, and an open text field where subject could give us their feedback freely. The first question asked the participant if he or she feels this type of demonstration promotes learning and memory compared to a conventional conference. Answer was given on a 5-point Likert scale ranging from 1 (“not at all”) to 5 (“a lot”). The second question asked the participant whether it was more entertaining than a traditional conference or course. Answer was given on a 5-point Likert scale ranging from 1 (“less entertaining”) to 5 (“more entertaining”). Question three addressed whether participant was more or less absorbed by the presentation on the 3D dome compared to a presentation on a conventional rectangular screen. Answer was given on a 5-point Likert scale ranging from 1 (“less absorbed”) to 5 (“more absorbed”). Finally, the fourth question asked if the participant felt discomfort (i.e., if he felt dizzy) due to the presentation on the 3D dome. Answer was given on a 5-point Likert scale ranging from 1 (“not at all”) to 5 (“a lot”). We collected data on a total of 134 participants in two distinct performance places, during four different days. 52 participants were men and 82 were woman with an average age of 30.4±17.3 years old across all participants (minimum age was 7; maximum age was 80).

Results from the questionnaire are shown in Figure 5. Our results show improved subjective learning compared to conventional methods, improved entertainment value, improved absorption into the material being presented, and little discomfort with no participant experiencing strong discomfort.

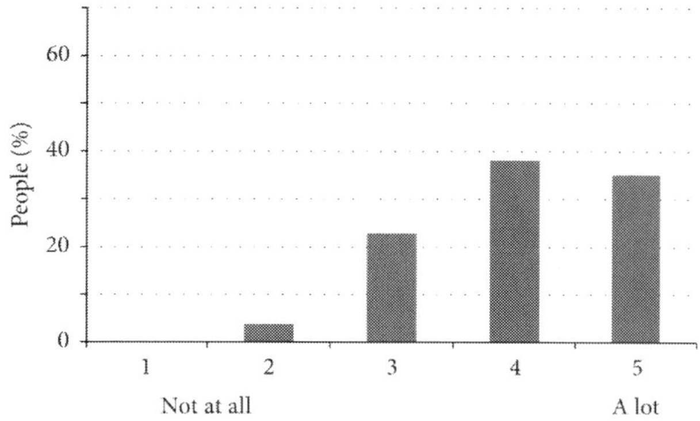
60
40
20
0
People (%)
1
2
3
4
5
Not at all
A lot

(a)

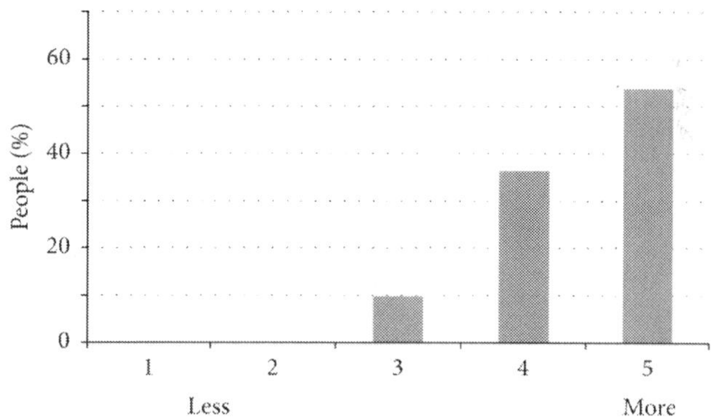
60
40
20
0
People (%)
1
2
3
4
5
Less
More

(b)

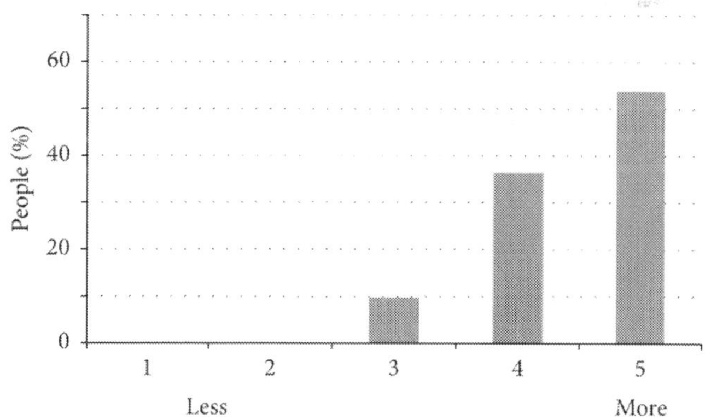
60
40
20
0
People (%)
1
2
3
4
5
Less
More

(c)

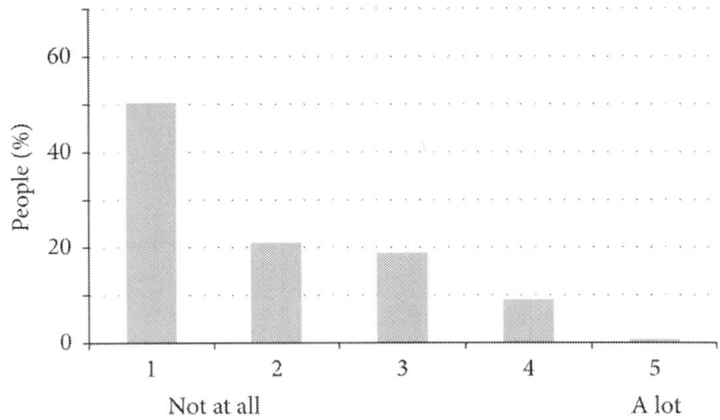

(d)

Figure 5: (a) Improved learning. (b) Entertainment. (c) Absorption. (d) Discomfort.

DISCUSSION

Planetarium domes have previously been used to display various contents. However, to our knowledge, this is the first time that real time EEG data is being shown in such an environment. Our demonstration appeared to arouse some level of popular success and seemed to provide participant with a new type of interactive experience. Thus, we have made all the tools we developed available in the public domain for anyone interested in reproducing our demonstration.

In the following sections, we will focus on four domains of application in which the Brainarium may potentially be used and further developed: education applications, entertainment applications, art applications, and immersive neurofeedback applications.

Education and Training Applications

The current Brainarium set-up already provides educational material to explain some basic concepts in Cognitive Sciences. We are currently exploring the possibility of showing content using stereoscopic projection methods, with the goal of providing an even more intense immersive experience to the public.

We currently focus on porting two classical BCI applications to the dome environment and developing pedagogical materials. The first application involves visualizing brain electrical activity related to emotion. Recent studies have reported that it is possible to differentiate emotional reactions and states using EEG in real time [14, 37]. When the participant wearing the EEG headset is experiencing a given emotion, an appropriate dynamical pattern reflecting the subject's emotion would be shown on the dome. The second application involves visualizing brain electrical activity associated with real and imagined body movements. Execution or mental visualization of body movement gives rise to typical brain rhythms [38]. These rhythms are recorded at the scalp surface and may be used to control visual display or even robotic devices. Moreover, results brought by fMRI studies on these domains can be shown to complement the explanations, showing brain areas and brain processes involved. The ultimate goal is to use the interactive and immersive dimensions to create and stimulate curiosity, attention, and interest in order to serve pedagogical purposes.

Entertainment Applications

The Brainarium could potentially be used as an immersive environment for BCI based games. BCI appear as a potential new way to gain control over a video game or a virtual world [39, 40]. Several EEG products specifically developed for BCI games have recently been made available to the general public in the form of commercial games (Star Wars Force Trainer and Mindflex by Matel, Inc.) and video games (Mindout: http://www.mindoutgame.com/, Free [41]). Several game studios have even specialized to solely design BCI games (MindGames: http://mindgames.is/, Dreams of Danu: http://www.dreamsofdanu.com/).

Immersive environments such as hemispheric projection surfaces have been already used for video games (e.g., with Blender full dome compatible Game Engine) [30], but never in conjunction with BCI systems. Moreover, it has been pointed out in a previous study by Lalor et al. [42] that subjects report that the multimodal feedback, such as the visuoauditive feedback delivered by the Brainarium, is useful in learning to control the game by suggesting that immersion increases sensation and therefore provides a more enjoyable game experience.

However, the engagement in the task of controlling the game using brainwaves might be too demanding and might degrade game experience. Nelson et al. [43] showed that concentration on the BCI task interacted with the feelings of presence in a virtual reality environment. However, they report as well that over time BCI control became more automatic for subjects as their brain adapts to the device, which allowed them to be gradually more absorbed by the virtual reality environment and feel more present. This description varies from what most subjects who experience the same virtual environment without BCI report: initially participants feel a high sense of presence which gradually drops as they realize the limitations of the virtual environment [44].

But what does the dome bring compared to a classic head mounted virtual reality device such as 3D goggles? An experiment studied the experience of users in an immersive device called the Cave [45], a room in which the user is presented with high-resolution stereo-pair images projected in real time on 3 walls and the floor, which provides an experience similar to a dome environment. They compared the experience of users in several environments: no immersion, head mounted 3D goggles, and the Cave. Subjects rated the Cave as providing a more immersive experience than all other conditions. Subjects also reported that the Cave was more comfortable than the head mounted goggles. There are numerous potential causes of visual discomfort when viewing stereo displays [46]. One of them is the vergence-accommodation conflict, that is, small amounts of left/right asymmetries, which is potentially present in all conventional stereo [47]. These results argue in favour of dome or room based systems for producing highly immersive environments.

Art Applications

More and more exploratory work using digital media and interactive devices are emerging on the art scene, leading to the relatively new field of interactive art. This developing genre of art usually has the public providing input in order to determine some parts or characteristics of the created content. Interactive art provides a ground for dialogue between the artist and the public through the potential of actions or reactions, introducing either intentional or passive ways to act upon the artwork.

The Brainarium is specific in the sense that the participant brainwaves are the source of interactivity. The artist may modulate multimedia artwork projected on the dome based on participant brainwaves. As mentioned for the education application parts, the artist may be able to extract subject's emotion and adapt the art forms being shown on the dome. Our system finally opens up the possibility to live coparticipation involving one or several participants wearing EEG headsets.

Medical Applications Using Immersive Neurofeedback

Neurofeedback is a type of brain computer interface application used in clinical environments to help to treat pathological traits [48–51]. Neurofeedback is being used to treat neuropsychological pathologies, epilepsy, ADHD, addiction, and depression [6, 52, 53], and to improve performance (stress management, creativity, attention and focus, and control of impulsivity [7, 8, 54–57]). The idea behind neurofeedback is that pathological mental states generate abnormal brain rhythms. By training patients to control their brain rhythms and suppress the pathological ones, it might be possible to treat specific mental pathologies. Note that neurofeedback is not yet widely accepted in the scientific and medical communities although recent neuroscientific works indicate some level of clinical efficacy and a bright future for this discipline [58–62].

Recent research results brought evidence that, in the context of neurofeedback training, immersion tends to improve training efficiency compared to classic feedback on a 2D screen [63]. As stated by Lécuyer et al. [64], virtual reality (VR) technologies provide motivating, safe, and controlled conditions that enable improvement of BCI learning. As reported in a recent review by Pfurtsheller et al. [65], a realistic virtual and immersive environment enhances the feeling of presence, task performance, and also cortical activation [66–68]. Studies indicate that the more game-like and engaging neurofeedback applications often resulted in a better performance [69, 70]. Subjects report the games are more stimulating and that multimodal immersive feedback is useful [42].

Previous studies have used virtual reality goggles with neurofeedback [63] but neurofeedback has never been performed in immersive environments like the one we are presenting here. Immersive environments could potentially offer numerous other benefits to patients, such as reduced training time, improved classification accuracy, increased sense of immersion and presence in an artificial setting, and reduced boredom or fatigue [71]. Finally, in the context of a therapeutic neurofeedback session, the dome environment provides a unique environment for enhanced intimacy between the patient and the therapist.

In the specific field of emotion regulation, fMRI neurofeedback recently brought very promising results [72–76]. However and despite the difficulty of recording subcortical regions of the brain involved in emotion generation, result obtained with EEG recordings [14, 15, 37] could be extended and refined in order to benefit from the high temporal resolution of the EEG and target in particular cortical areas involved in emotion monitoring and regulation [77]. Independent component analysis and source reconstruction methods could potentially be used to improve EEG spatial resolution and signal to noise ratio. Cannon et al. [78] showed that limbic lobe and hippocampal activity can be recorded and visualized using LORETA during affective memory recall. In another study, Cannon et al. [79] showed that it was possible to learn to self-regulate activity in anterior cingulated gyrus, an area of the brain known to be involved in both cognitive and affective processes. ICA neurofeedback and LORETA neurofeedback are indeed possible in an immersive set-up such as the Brainarium. Following recent developments in the field of virtual reality technology, several studies argued in favour of several benefits from using virtual reality in treatment of various pathologies or disorders related to emotions such as anxiety disorders (for a review see [80]). Bringing together BCI and VR could potentially help to not only better monitor and therefore optimize the therapy, but also give birth to new therapeutic techniques.

CONCLUSION

We described the first interactive system allowing real time spatialized visualization of electrical brain activity in a brain-like shaped immersive environment. This device was initially intended to deliver scientific knowledge using a pedagogical medium at the crossing between art, science, and technology. Its modular architecture allows extending and adapting it to various implementation solutions leveraging the costs to different contexts of deployment. This innovative concept can be further developed into a rich variety of applications in educational, entertainment, art, and medical domains.

Competing Interests

The authors declare that they have no competing interests.

Acknowledgments

The authors wish to thank the Art and Science Association "Les Chemins Buissonniers" that collaborated on this project, the association "La Ligue de l'Enseignement" that lent the dome for the demos, the association "InCOGnu" who provided the Emotiv Epoc headset, and Isabelle Cirla, the professional musician who played during the live exhibits. This project was supported by a grant in neuroinformatics from the FRM Foundation and a grant from the French Ministry of Research on brain computer interfaces.

REFERENCES

1. C. S. Sherrington, The Integrative Action of the Nervous System, Yale University Press, New Haven, Conn, USA, 1906.
2. H. R. Maturana and F. J. Varela, The Tree of Knowledge: The Biological Roots of Human Understanding, New Science Library/ Shambhala Publications, 1987.
3. J. R. Wolpaw, N. Birbaumer, D. J. McFarland, G. Pfurtscheller, and T. M. Vaughan, "Brain-computer interfaces for communication and control," Clinical Neurophysiology, vol. 113, no. 6, pp. 767–791, 2002. · ·
4. B. Z. Allison, E. W. Wolpaw, and J. R. Wolpaw, "Brain-computer interface systems: progress and prospects," Expert Review of Medical Devices, vol. 4, no. 4, pp. 463–474, 2007.
5. C. S. Nam, G. Schalk, and M. M. Jackson, "Current trends in Brain-Computer Interface (BCI) research and development," Journal of Human-Computer Interaction, vol. 27, no. 1, pp. 1–4, 2010. ·
6. J. Gruzelier and T. Egner, "Critical validation studies of neurofeedback," Child and Adolescent Psychiatric Clinics of North America, vol. 14, no. 1, pp. 83–104, 2005. · ·
7. J. Gruzelier, T. Egner, and D. Vernon, "Validating the efficacy of neurofeedback for optimising performance," Progress in Brain Research, vol. 159, pp. 421–431, 2006. ·

8. T. Thompson, T. Steffert, T. Ros, J. Leach, and J. Gruzelier, "EEG applications for sport and performance," Methods, vol. 45, no. 4, pp. 279–288, 2008.
9. H. Berger, "Über das Elektrenkephalogramm des Menschen," Archiv für Psychiatrie und Nervenkrankheiten, vol. 87, no. 1, pp. 527–570, 1929.
10. E. D. Adrian and B. H. C. Matthews, "The berger rhythm: potential changes from the occipital lobes in man," Brain, vol. 57, no. 4, pp. 355–385, 1934.
11. G. Pfurtscheller, C. Neuper, and W. Mohl, "Event-related desynchronization (ERD) during visual processing," International Journal of Psychophysiology, vol. 16, no. 2-3, pp. 147–153, 1994.
12. A. Lucier, "Statement on: music for solo performer," in Biofeedback and the Arts, Results of Early Experiments, D. Rosenboom, Ed., pp. 60–61, Aesthetic Research Center of Canada Publications, Vancouver, Canada, 1976.
13. J. Kamiya, "Operant control of the EEG alpha rhythm and some of its reported effects on consciousness," in Altered States of Consciousness, pp. 519–529, John Wiley & Sons, New York, NY, USA, 1969.
14. Y.-P. Lin, J.-R. Duann, J.-H. Chen, and T.-P. Jung, "Electroencephalographic dynamics of musical emotion perception revealed by independent spectral components," NeuroReport, vol. 21, no. 6, pp. 410–415, 2010.
15. G. Leslie, A. Ojeda, and S. Makeig, "Towards an affective brain-computer interface monitoring musical engagement," in Proceedings of the 5th Humaine Association Conference on Affective Computing and Intelligent Interaction (ACII '13), pp. 871–875, Geneva, Switzerland, September 2013.
16. S. Mann, J. Fung, and A. Garten, "DECONcert: bathing in the light, sound, and waters of the musical brainbaths," 2007.
17. G. Leslie and T. Mullen, "MoodMixer: EEG-based Collaborative Sonification," in Proceedings of the International Conference on New Interfaces for Musical Expression, Oslo, Norway, June 2011.
18. T. Mullen, R. Warp, and A. Jansch, "Minding the (transatlantic) gap: an internet-enabled acoustic brain-computer music interface,"

in Proceedings of the International Conference on New Interfaces for Musical Expression, Oslo, Norway, May-June 2011.

19. E. R. Miranda and A. Brouse, "Interfacing the brain directly with musical systems: on developing systems for making music with brain signals," Leonardo, vol. 38, no. 4, pp. 331–336, 2005.
20. J. Clarke, Spectacle of the Mind, Global Mind Project, 2009, http://www.academia.edu/199451/Spectacle_of_the_Mind.
21. Tkoranyi, Xxxy's Infinity Simulator Connects Emotiv Headset to Theatrical Flying Rig, Neurogadget, 2011.
22. Digitalis Education Solutions, "Digitarium Delta Digital Planetarium System," In: Digitalis Education Solutions, 2011, http://www.digitaliseducation.com/products-digitarium_delta.html.
23. Digitalis Education Solutions, Digitalis Portable Planetarium Domes, 2011, http://www.digitaliseducation.com/domes.html.
24. Y. Lhoumeau and L. Ruiz, "Lhoumeau Sky-System (Numeric Planetarium with Fisheye)," 2010, http://www.lss-planetariums.info/index.php.
25. P. Bourke, "Digital fulldome, techniques and technologies," in Proceedings of the Graphite Conference, ACM SIGGRAPH, Perth, Australia, December 2007.
26. P. Bourke, "Using a spherical mirror for projection into immersive environments," in Proceedings of the 3rd International Conference on Computer Graphics and Interactive Techniques in Australasia and South East Asia, pp. 281–284, Dunedin, New Zealand, 2005.
27. Emotiv Inc, "Emotiv EPOC Software Development Kit," Emotiv—Brain Computer Interface Technology, 2010, https://github.com/Emotiv.
28. M. Wright, "OpenSoundControl," in OpenSound Control, 2011, http://archive.cnmat.berkeley.edu/OpenSoundControl/.
29. Y. Renard, F. Lotte, G. Gibert et al., "OpenViBE: an open-source software platform to design, test, and use brain-computer interfaces in real and virtual environments," Presence: Teleoperators and Virtual Environments, vol. 19, no. 1, pp. 35–53, 2010.
30. P. Bourke and D. Felinto, "Blender and immersive gaming in a hemispherical dome," GSTF International Journal on Computing, vol. 1, no. 1, 2010.

31. R. Grandchamp, "Downloads," Romain Grandchamp's Homepage, 2011, https://sites.google.com/site/romaingrandchamp/cerveaurium/downloads.

32. E. Pereda, A. Gamundi, R. Rial, and J. González, "Non-linear behaviour of human EEG: fractal exponent versus correlation dimension in awake and sleep stages," Neuroscience Letters, vol. 250, no. 2, pp. 91–94, 1998.

33. R. Ferenets, T. Lipping, A. Anier, V. Jäntti, S. Melto, and S. Hovilehto, "Comparison of entropy and complexity measures for the assessment of depth of sedation," IEEE Transactions on Biomedical Engineering, vol. 53, no. 6, pp. 1067–1077, 2006.

34. D. S. Bassett, A. Meyer-Lindenberg, S. Achard, T. Duke, and E. Bullmore, "Adaptive reconfiguration of fractal small-world human brain functional networks," Proceedings of the National Academy of Sciences of the United States of America, vol. 103, no. 51, pp. 19518–19523, 2006.

35. P. A. Carpenter, C. J. Davia, and R. L. Vimal, "Catalysis, perception, and consciousness," New Mathematics and Natural Computation, vol. 5, no. 1, pp. 287–306, 2009.

36. B. B. Mandelbrot, "Fractal aspects of the iteration of $z \rightarrow \lambda z\,(1-z)$ for complex λ and z," Annals of the New York Academy of Sciences, vol. 357, no. 1, pp. 249–259, 1980.

37. Y.-P. Lin, C.-H. Wang, T.-L. Wu, S.-K. Jeng, and J.-H. Chen, "EEG-based emotion recognition in music listening: a comparison of schemes for multiclass support vector machine," in Proceedings of the 2009 IEEE International Conference on Acoustics, Speech, and Signal Processing (ICASSP ‹09), pp. 489–492, IEEE, Taipei, Taiwan, April 2009.

38. G. Pfurtscheller and F. H. Lopes da Silva, Event-Related Desynchronization, Elsevier Health Sciences, Philadelphia, Pa, USA, 1999.

39. B. Reuderink, "Games and brain-computer interfaces: the state of the art," WP2 BrainGain Deliverable, HMI, University of Twente, Enschede, The Netherlands, 2008.

40. A. Nijholt, D. P.-O. Bos, and B. Reuderink, "Turning shortcomings into challenges: brain-computer interfaces for games," Entertainment

Computing, vol. 1, no. 2, pp. 85–94, 2009.

41. Emotiv, Emotiv Application Store. In: Emotiv—Brain Computer Interface Technology, 2010, http://emotiv.com/product-category/applications/.

42. E. C. Lalor, S. P. Kelly, C. Finucane et al., "Steady-state VEP-based brain-computer interface control in an immersive 3D gaming environment," EURASIP Journal on Applied Signal Processing, vol. 2005, no. 19, pp. 3156–3164, 2005.

43. W. T. Nelson, L. J. Hettinger, J. A. Cunningham, M. M. Roe, M. W. Haas, and L. B. Dennis, "Navigating through virtual flight environments using brain-body-actuated control," in Proceedings of the IEEE Virtual Reality Annual International Symposium, pp. 30–37, IEEE Computer Society, Albuquerque, NM, USA, March 1997. ·

44. M. Garau, H. R. Widenfeld, A. Antley et al., "Temporal and spatial variations in presence: a qualitative analysis," in Proceedings of the International Workshop on Presence, pp. 232–239, Valencia, Spain, January 2004.

45. C. Cruz-Neira, J. Leigh, M. Papka et al., "Scientists in wonderland: a report on visualization applications in the CAVE virtual reality environment," in Proceedings of the IEEE Symposium on Research Frontiers in Virtual Reality, pp. 59–66, San Jose, Calif, USA, October 1993. ·

46. F. L. Kooi and A. Toet, "Visual comfort of binocular and 3D displays," Displays, vol. 25, no. 2-3, pp. 99–108, 2004.

47. T. Shibata, J. Kim, D. M. Hoffman, and M. S. Banks, "The zone of comfort: predicting visual discomfort with stereo displays," Journal of Vision, vol. 11, no. 8, article 11, 2011.

48. N. Birbaumer and L. G. Cohen, "Brain-computer interfaces: communication and restoration of movement in paralysis," The Journal of Physiology, vol. 579, no. 3, pp. 621–636, 2007.

49. E. Buch, C. Weber, L. G. Cohen et al., "Think to move: a neuromagnetic brain-computer interface (BCI) system for chronic stroke," Stroke, vol. 39, no. 3, pp. 910–917, 2008.

50. N. Birbaumer, A. Ramos Murguialday, C. Weber, and P. Montoya, "Chapter 8 Neurofeedback and Brain-Computer Interface. Clinical

Applications," International Review of Neurobiology, vol. 86, pp. 107–117, 2009.

51. W. Wang, J. L. Collinger, M. A. Perez et al., "Neural interface technology for rehabilitation: exploiting and promoting neuroplasticity," Physical Medicine and Rehabilitation Clinics of North America, vol. 21, no. 1, pp. 157–178, 2010.
52. H. Heinrich, H. Gevensleben, and U. Strehl, "Annotation: neurofeedback—train your brain to train behaviour," Journal of Child Psychology and Psychiatry and Allied Disciplines, vol. 48, no. 1, pp. 3–16, 2007.
53. M. B. Sterman, "Basic concepts and clinical findings in the treatment of seizure disorders with EEG operant conditioning," Clinical EEG Electroencephalography, vol. 31, no. 1, pp. 45–55, 2000.
54. E. Angelakis, S. Stathopoulou, J. L. Frymiare, D. L. Green, J. F. Lubar, and J. Kounios, "EEG neurofeedback: a brief overview and an example of peak alpha frequency training for cognitive enhancement in the elderly," The Clinical Neuropsychologist, vol. 21, no. 1, pp. 110–129, 2007.
55. T. Egner and J. H. Gruzelier, "Learned self-regulation of EEG frequency components affects attention and event-related brain potentials in humans," NeuroReport, vol. 12, no. 18, pp. 4155–4159, 2001.
56. B. Zoefel, R. J. Huster, and C. S. Herrmann, "Neurofeedback training of the upper alpha frequency band in EEG improves cognitive performance," NeuroImage, vol. 54, no. 2, pp. 1427–1431, 2011.
57. P. Wiener, "Neurotherapy in adult management of stress: developmental aspects of the adult immune system," Journal of Adult Development, vol. 18, no. 2, pp. 66–69, 2011.
58. H. Gevensleben, B. Holl, B. Albrecht et al., "Distinct EEG effects related to neurofeedback training in children with ADHD: a randomized controlled trial," International Journal of Psychophysiology, vol. 74, no. 2, pp. 149–157, 2009.
59. J. P. Hamilton, G. H. Glover, J.-J. Hsu, R. F. Johnson, and I. H. Gotlib, "Modulation of subgenual anterior cingulate cortex activity with real-time neurofeedback," Human Brain Mapping, vol. 32, no. 1, pp. 22–31, 2011.

60. I. H. C. H. M. Philippens and R. A. P. Vanwersch, "Neurofeedback training on sensorimotor rhythmin marmoset monkeys," NeuroReport, vol. 21, no. 5, pp. 328–332, 2010.
61. S. Wangler, H. Gevensleben, B. Albrecht et al., "Neurofeedback in children with ADHD: specific event-related potential findings of a randomized controlled trial," Clinical Neurophysiology, vol. 122, no. 5, pp. 942–950, 2011.
62. R. J. Schafer and T. Moore, "Selective attention from voluntary control of neurons in prefrontal cortex," Science, vol. 332, no. 6037, pp. 1568–1571, 2011.
63. J. Gruzelier, A. Inoue, R. Smart, A. Steed, and T. Steffert, "Acting performance and flow state enhanced with sensory-motor rhythm neurofeedback comparing ecologically valid immersive VR and training screen scenarios," Neuroscience Letters, vol. 480, no. 2, pp. 112–116, 2010.
64. A. Lécuyer, F. Lotte, R. B. Reilly, R. Leeb, M. Hirose, and M. Slater, "Brain-computer interfaces, virtual reality, and videogames," Computer, vol. 41, no. 10, pp. 66–72, 2008.
65. G. Pfurtscheller, R. Leeb, J. Faller, and C. Neuper, "Brain-computer interface systems used for virtual reality control," in Virtual Reality, J.-J. Kim, Ed., chapter 1, pp. 3–20, InTech, Rijeka, Croatia, 2011.
66. L. Jäncke, M. Cheetham, and T. Baumgartner, "Virtual reality and the role of the prefrontal cortex in adults and children," Frontiers in Neuroscience, vol. 3, pp. 52–59, 2009.
67. S. Lee and G. J. Kim, "Effects of visual cues and sustained attention on spatial presence in virtual environments based on spatial and object distinction," Interacting with Computers, vol. 20, no. 4-5, pp. 491–502, 2008.
68. M. Slater, A. Steed, and Y. Chrysanthou, Computer Graphics and Virtual Environments: From Realism to Real-Time, Addison Wesley, 2002.
69. J. D. Bayliss, S. A. Inverso, and A. Tentler, "Changing the P300 brain computer interface," Cyberpsychology and Behavior, vol. 7, no. 6, pp. 694–704, 2004.
70. D. Oude Bos and B. Reuderink, "BrainBasher: a BCI game," in Proceedings of the Extended Abstracts of the International

Conference on Fun and Games, vol. 4, pp. 36–39, Eindhoven, Netherlands, October 2008, http://doc.utwente.nl/65050/.

71. R. Ron-Angevin, M. A. Lopez, and F. Pelayo, "The training issue in brain-computer interface: a multi-disciplinary field," in Bio-Inspired Systems: Computational and Ambient Intelligence, J. Cabestany, F. Sandoval, A. Prieto, and J. M. Corchado, Eds., vol. 5517 of Lecture Notes in Computer Science, pp. 666–673, Springer, Berlin, Germany, 2009. ·
72. V. Zotev, R. Phillips, K. D. Young, W. C. Drevets, and J. Bodurka, "Prefrontal control of the amygdala during real-time fMRI neurofeedback training of emotion regulation," PLoS ONE, vol. 8, no. 11, Article ID e79184, 2013.
73. D. E. J. Linden, I. Habes, S. J. Johnston et al., "Real-time self-regulation of emotion networks in patients with depression," PLoS ONE, vol. 7, no. 6, Article ID e38115, 2012. ·
74. V. Zotev, F. Krueger, R. Phillips et al., "Self-regulation of amygdala activation using real-time fMRI neurofeedback," PLoS ONE, vol. 6, no. 9, Article ID e24522, 2011.
75. A. Caria, R. Sitaram, R. Veit, C. Begliomini, and N. Birbaumer, "Volitional control of anterior insula activity modulates the response to aversive stimuli. A real-time functional magnetic resonance imaging study," Biological Psychiatry, vol. 68, no. 5, pp. 425–432, 2010.
76. S. J. Johnston, S. G. Boehm, D. Healy, R. Goebel, and D. E. J. Linden, "Neurofeedback: a promising tool for the self-regulation of emotion networks," NeuroImage, vol. 49, no. 1, pp. 1066–1072, 2010.
77. K. N. Ochsner and J. J. Gross, "The cognitive control of emotion," Trends in Cognitive Sciences, vol. 9, no. 5, pp. 242–249, 2005.
78. R. Cannon, J. Lubar, K. Thornton, S. Wilson, and M. Congedo, "Limbic beta activation and LORETA: can hippocampal and related limbic activity be recorded and changes visualized using LORETA in an affective memory condition?" Journal of Neurotherapy, vol. 8, no. 4, pp. 5–24, 2004.

79. R. Cannon, J. Lubar, A. Gerke, K. Thornton, T. A. Hutchens, and V. McCammon, "EEG spectral-power and coherence: LORETA neurofeedback training in the anterior cingulate gyrus," Journal of Neurotherapy, vol. 10, no. 1, pp. 5–31, 2006.

80. A. Gorini and G. Riva, "Virtual reality in anxiety disorders: the past and the future," Expert Review of Neurotherapeutics, vol. 8, no. 2, pp. 215–233, 2008.

CHAPTER 7

CREATIVITY AS CENTRAL TO CRITICAL REASONING AND THE FACILITATIVE ROLE OF MORAL EDUCATION: UTILIZING INSIGHTS FROM NEUROSCIENCE

Terence Lovat[1], Daniel Fleming[2]

[1]The University of Newcastle, Newcastle, Australia

[2]The Broken Bay Institute, Sydney, Australia

ABSTRACT

The article will review the literature of updated neuroscientific research in order to gain insights into the centrality to critical reasoning of creativity (understood as creative thinking, including the impulsion of imagination and wonder). Furthermore, it will explore literature that testifies to the credentials of moral education in facilitating the forms of creativity associated with the development of critical reasoning, as suggested by the insights of neuroscience. Finally, the article will review the literature related to earlier empirical evidence of the crucial role that moral education can play in facilitating creativity, imagination and wonder.

Keywords: Creativity, Wonder, Imagination, Critical Reasoning, Moral Education, Neuroscience

INTRODUCTION

The study design is a literature review aimed at eliciting evidence of new directions in education concerned with the central role of creativity, wonder and imagination in facilitating critical reasoning of the kind that impels educational achievement. Furthermore, it appraises literature that illustrates the potential of moral education to contribute to the development of creativity, wonder and imagination and to empirical evidence of this development. It begins by contextualizing current educational thinking that points to the creativity/critical reasoning link as evidenced in recent documentation in Australian education.

LINKING CREATIVITY AND CRITICAL REASONING IN AUSTRALIAN EDUCATIONAL DOCUMENTATION

As an instance of new and broader directions in holistic thought about learning, the Melbourne Declaration on Educational Goals for Young Australians (MCEETYA, 2008) identifies "creativity" among the essential skills for twenty-first century learners. An allied, supplementary document (ACARA, 2013), designed to outline the general capabilities needed to achieve the above goals, specifies the following:

Responding to the challenges of the twenty-first century—with its complex environmental, social and economic pressures—requires young people to be creative, innovative, enterprising and adaptable, with the motivation, confidence and skills to use critical and creative thinking purposefully (p. 66).

In underlining what it proposes to be an inherent link between creativity and critical reasoning,ACARA (2013) proffers:

Critical and creative thinking are integral to activities that require students to think broadly and deeply using skills, behaviours and dispositions such as reason, logic, resourcefulness, imagination and innovation in all learning areas at school and in their lives beyond school … Creative thinking involves students in learning to generate and apply new ideas in specific contexts, seeing existing situations in a new way, identifying alternative explanations, and seeing or making new links that generate a positive outcome ... Dispositions such as inquisitiveness, reasonableness, intellectual flexibility, open- and fair-mindedness, a readiness to try new ways of doing things and consider alternatives, and persistence both promote and are enhanced by critical and creative thinking ... Critical and creative thinking can be encouraged simultaneously through activities that integrate reason, logic, imagination and innovation (pp. 66-67).

This Australian work is reflective of much attention in current educational scholarship to the links between creativity and critical reasoning. Such scholarship includes literature from a number of sources, as identified below. Higgins-D'Alessandro (2011) highlights the interconnections between rational, critical and imaginative thinking. Bryk (2012) underlines the need for educational policy that nurtures human creativity as a feature of the kind of reasoning required for effective learning.Newmann (2013) , under the title "Muzzling Minds all over the Globe while Trumpeting Higher Order Skills", underlines the centrality of creativity and imagination to the most efficacious forms of higher order thinking (read "critical reasoning"). Scholarship that makes these connections has been strengthened by updated findings emanating from the neurosciences and the paper will now appraise an array of updated extant literature that strengthens the creativity/critical reasoning linkage.

NEUROSCIENCE: LITERARY EVIDENCE OF THE CENTRALITY OF CREATIVITY

The findings of neuroscience have become more certain with the advent of advanced technology capable of tracing and digitally representing activity within and among the fibres of the brain (Srinivasan, 2007 ; Jensen, 2008 ; Raichle, 2009 ; Friston et al., 2011 ; Meltzoff et al., 2009). These images confirm the nexus between activity more traditionally understood as cognition and the array of activities concerned with a range of other neural responses (Narvaez, 2010). Furthermore, these findings have been brought into a direct relationship with matters concerned with education, be it in its foundations, pedagogy or curriculum (Varma et al., 2008 ; Rosiek & Beghetto, 2009 ; Immordino-Yang & Faeth,

2010 ; Dommett et al., 2011 ; Hille, 2011 ;Immordino-Yang, 2011 ; Lovat et al., 2011 ; Schrag, 2011 ; Clement & Lovat, 2012).

Within the array of neuroscientific data, there are particular elements that highlight the crucial nature of creativity for normal brain development, and most especially as a necessary facilitator of the type of critical reasoning ability most essential to effective learning. Doidge (2009) asserts that the findings of updated neuroscience have shifted fundamentally our conception of what the brain is and how it functions from one that sees it as 'hardwired' to one that is "plastic". By this, he means that we now know the cells and fibres in the brain to be so malleable as to be readily amenable to both external and internal influences. Experiences of joy, grief, love, hate, encouragement and discouragement will all impact on the way the brain functions and, over extended periods, how the brain grows and develops—or not. Similarly, Sajjadi (2008) argues that recent neuroscientific evidence renders the traditional notion of the "brain as machine" as quite inadequate if education is to optimize the potential for learning. Rather than seeing the brain as mechanistic and arborescent (or "tree-like"), we need to work on the conception that it is dynamic and rhizomatic (or "root-like"). Instead of the seemingly static tree, the brain is more like the ever-growing and expanding roots, influenced by the experiences to which the growing person is subjected, especially in the early years of development (Narvaez, 2013).

According to Narvaez (2010, 2013) , the ramifications for educators rest with their capacity to impact on the brain's development through the kinds of content, curriculum and pedagogy they employ. Furthermore, the brain is so complex, with every element impacting on every other element, that learning must be a multivariate and enriching experience for students (Immordino-Yang & Faeth, 2010). If learning challenges are all premised on the notion of the brain as "hardwired" and "mechanistic", being simply able to receive information, process and regurgitate it as required, then education has the potential to be a negative and even damaging process for students, especially for those not so well equipped to deal with simple reception, processing and regurgitation (Immordino-Yang & Damasio, 2007). Not only can their self-esteem be damaged but, according to the assumptions of neuroplasticity, their brains can fail to grow in line with their potential (Doidge, 2009).

Narvaez (2008, 2010, 2013, 2014) takes the case further in the pivotal role she assigns to creativity as providing for the emotional stability that is essential to sound and critical reasoning. Herein, she employs updated neuroscientific evidence around the role of the pre-frontal cortex of the human brain, showing it to be its most evolved and distinctive feature. Especially in the context of a caring and nurturing environment, neural activity associated with imagination is the key to establishing the emotional regulation necessary to the forming of effective decision-making and problem-solving. In turn, the self-confidence that results from such decision-making and problem-solving becomes the basis of ongoing critical reasoning, and indeed of learning. A caring and nurturing environment is essential, but not sufficient, for this to occur. What is also needed is the guiding hand of efficacious pedagogy. Narvaez makes use of McGilchrist (2009) in identifying with some precision how the right and left hemispheres of the brain function, the left being more associated with rationality and the right with affectivity. Narvaez (2013) clearly, and perhaps surprisingly, associates creativity principally with left brain activity, with the neural capacity to abstract, reflect, self-reflect and problem solve. It is, in this sense, the key to unlocking the potential of critical reasoning to drive new learning. In her most recent work, Narvaez (2014) also links this aspect of creativity to a broader vision of flourishing for the human person, claiming that while modern, mostly Western, understandings of creativity have tended to concentrate on 'detached imagination focused on technical manipulation and innovation (p. 245). Narvaez claims that a more holistic understanding of human flourishing, which emphasises humankind's relatedness to the natural world, along with moral values such as generosity and compassion, is better positioned to allow for the full functioning of the human brain and, correlatively, its capacities for creativity.

We conclude this section of the literature review by referring to an array of prominent literature that captures the healthy scepticism that exists among educational researchers about the capacity and/or appropriateness of neuroscience being applied to education. Snook (2012) proposes that he is not opposed to scientific appraisal of education and that elements of neuroscience may well have something to offer but that we need to differentiate "good science" from fads and "neuromyths", among which he includes the notion of "left and right brain thinking". Weisberg et al. (2008)confirm the potential mischievousness of "neuromythology" through an empirical study that illustrated the "allure" of reference to neuroscience, even when irrelevant to the issue at hand, on the perceptions of study subjects, most especially non-expert subjects. Bruer (1997, 1998, 2006)has gained a reputation as an arch-sceptic of neuroscience's relevance to education, arguing that the applications made are too often over-simplified links from biological evidence to the very different field of teaching and learning. He argues that the missing link in effective application is psychology and that, without its filter, many of the arguments proffered for neuroscience's implications for the practicalities of education are unfounded.

This authorship remains aware of the debate about and the contentiousness about the limitations of neuroscience for educational application (Clement & Lovat, 2012) but, importantly also notes the many neuroscientists and educational psychologists, as illustrated above, who argue for and have achieved reputable empirical studies that seem to confirm the applicability of neuroscience's insights to education. This authorship makes no claim to scientific expertise but has proffered its own view that, regardless of the scientific argumentation one way or the other, the insights of neuroscience can be useful as metaphorical devices for analysing educational practices (Fleming & Lovat, 2015 in press). Furthermore, it has written refereed work, including elements of its own cmpirical work (Fleming & Lovat, 2014) that illustrates how such devices can be useful in education, especially in negotiating teaching and learning in challenging multicultural, multi-value contexts. In a word, we believe the authorship has demonstrated the creative potential of neuroscience insights being applied to educational thinking in a way that is particularly pertinent to the purpose of this article.

CREATIVITY, IMAGINATION AND WONDER

The article now expands on the literature review to incorporate a wider array of scholarship that has made use of neuroscience insights in proffering the crucial nature of imagination and wonder, as artefacts and extensions of creativity, in developing critical reasoning. While the three capacities connoted by creativity, imagination and wonder might be regarded as either insufficiently alike to be lumped together or so alike as to be unnecessarily separately named, we argue that the abovementioned updated neuroscientific evidence on which we rely shows that, on the one hand, they complement and complete each other in a particularly rich way but, on the other hand, that they each denote a facet of the evidence that justifies their being seen as synergistic but separate skills.

Jordan and Carlisle (2013) refer to creativity as "a conceptual ability to come up with new ideas that are surprising, yet intelligible" (p. 7). The earlier work of Dietrich (2004) utilized updated neuroscientific evidence in forging the links between creativity and intelligibility, showing that, especially when creativity is impelled by deliberate control (such as in a formal learning situation), it stimulates the pre-frontal cortex and, as identified above, it is the most developed and distinctively human part of the brain, the part on which critical reasoning relies most fully. Further research seems to confirm Dietrich's earlier work in this regard. Srinivasan (2007) employed EEG data in illustrating that creative tasks utilized the pre-frontal cortex, especially when related to "... effortful problem-solving tasks" (p. 109). In this sense, creativity can be seen to be a regular but vital component of intelligibility. Wegerif et al. (2010) went further in using artificial intelligence pattern matching techniques to show that creative activity could be coded and read by using more established critical thinking codes. Their work again underlines the synergy to be found in brain activity between creativity and critical reasoning.

In her 'Triune Ethics Theory' (TET), in essence a theory drawing on and directed to moral education, Narvaez (2008, 2013) also focuses on activity in the pre-frontal cortex necessary to the development of sound, critical reasoning, labouring the crucial part played by imaginative thinking in stimulating the left brain formation that is essential to the development of such reasoning. It could be argued that Narvaez might as well have used the language of "creativity", rather than "imagination", in this theory because the two words appear virtually synonymous. By utilizing imagination, however, Narvaez is able to explore a variant of neuroscientific research which delves even further into the workings of the brain and the similarities to be found between creativity and critical reasoning, while also making use of some of the more extended evidence of enriching by the former on the latter.

Imagination has been found to be a particularly important focus for neuroscientific research aimed at identifying the fine-tuned similarities and differences between creativity and intelligibility, as normally understood. Hassabis et al. (2007) , for instance, used advanced brain scanning technology to trace similarities and differences in pre-frontal cortical activity between imagined fictitious experiences and episodic memory. In terms of the essential commonality of scene construction, there were 'remarkable similarities' in the effect on the pre- frontal cortex. This finding in itself underlines an aspect of the synergy between creativity and intelligibility and offers clues to the educator about the vast and too often untapped potential for imaginative thinking to stimulate critical reasoning related to real-life experiences and dilemmas.

Hassabis et al. (2007) also point to the likelihood of other brain functions coming into play that discriminate between imagined fictitious experiences and episodic memory, a notion that adds even more value to imagination as an educational strategy, not least in an era that tends to privilege rationality and empiricism over creativity. Combining Hassabis et al. with Doidge's (2009)"rhizomatic" brain notion, it would seem that neuroscience is pointing to creativity as a distinctively powerful way in which the full functioning of the brain can be effected. Furthermore, we can see that through the use of the notion of imagination and imaginative thinking, this insight into creativity's potential can be strengthened. We will show below that, especially in the application of her thinking to moral education, Narvaez's use of imagination allows her to explore profound and unlikely dimensions of creativity, including potentially negative and quite destructive ones that would have stretched the normal meaning attached to the word and would go beyond the evidence, at least within the ambit of neuroscience.

Just as imagination adds value to the concept of creativity, as above, so the same can be said of "wonder". Bulkeley (2005) , in The Wondering Brain, employs updated neuroscience to speak of the full range of emotions attached to those creative moments when something entirely new happens to one's intelligibility, powers of insight or what we refer to simply as critical reasoning. Bulkeley chooses to speak in this context of "wonder", which she defines as "… feeling excited by an encounter with something novel and unexpected, something that strikes a person as intensely real, true and/or beautiful" (p. 3). She points out that wonder has often been associated with the religious and moral traditions but that experiences of wonder have been "… crucial but unappreciated inspirations for … scientific progress and technological innovations" (p. 3). Why is this so? Bulkeley asks, and then answers her own question in the following way:

> To feel wonder is to experience a sudden decentering of the self. Facing something surprisingly new and unexpectedly powerful, one's ordinary sense of personal identity is dramatically altered, leading to new knowledge and understanding that ultimately recenter the self. The profound impact … is evident in both the intense memorability of the experiences and the strong bodily sensations that often accompany them. People speak of being stunned, dazed, breath-taken, overwhelmed, consumed, astonished—all gesturing toward a mode of experience that exceeds ordinary language and thought and yet inspires a yearning to explore, understand and learn … where the powerful emotional experience (of wonder) stimulates lively curiosity, knowledge-seeking behaviour and critical questioning" (p. 4).

Bulkeley (2005) makes much use of updated neuroscientific research around pre-frontal cortical activity in making her case about wonder as both an integral dimension of intelligibility as well as possessing potential to disrupt, disturb, de-activate and re-activate regular brain functions in order to stimulate the brain to expand its capacity for reasoning:

> The large expanses of association cortex that distinguish the human brain are hyperactivated as radically new input must be processed, upsetting established neural systems and forcing the creation of new ones. In terms of the range and complexity of neural connections, I would propose as a testable hypothesis that wonder makes the brain grow… The capacity to experience wonder is itself a developmental achievement ... wonder as existential surprise becomes wonder as knowledge-seeking curiosity (pp. 198-199).

As above, we are conscious of the contentiousness that exists around the validity of applications from neuroscience to education and related matters, in the way Narvaez has done (see for example, Haidt, 2010). We are at the same time buoyed by the overall positive response from scholarship that her works (Narvaez et al., 2013; 2014) have received in formal academic reviews (Bielert & Gallup, 2014), as well as previous evaluations of her approach (Lewis, 2010). It is similarly the case with

Bulkeley's work which, in spite of the hearty debate about neuroscience's potential to inform about practical enterprises, has received very positive academic reviews (cf.Gortner, 2007). On the basis of such endorsement, we believe it is valid to continue to explore the insights of neuroscience in an endeavour to gain enhanced insight into the potential of creative, imaginative and wondrous thinking to impact on critical reasoning.

The case for creativity, imagination and wonder as complementary in the development of critical reasoning, yet beneficially separable in their emphases and orientations, rests on analyses such as those above. Furthermore, we see, especially in Narvaez and Bulkeley, important references to the moral dimension of the kind of creativity that has potential to facilitate critical reasoning. As such, we wish to explore literature that pertains to moral education as an area of learning with especially strong credentials in stimulating this kind of creativity.

THE ROLE OF MORAL EDUCATION IN CULTIVATING CREATIVITY

As suggested above, Narvaez's concept of imagination as vital to the development of left brain activity leading to sound or critical reasoning is to be found in its most comprehensive form in her "Triune Ethics Theory" (Narvaez, 2008, 2013, 2014), a theory about human mindsets premised heavily on the impulsions we refer to as moral motivation and the desired intentions and directions of the enterprise we call moral education. Bulkeley's (2005) work on wonder is also set in the context of knowledge-gathering as an inherently moral enterprise when its goal is to truly extend the mind beyond the known to the new and surprising, for Bulkeley the only goal with capacity to stretch the powers of the brain. One gains a strong sense from these perspectives that the earlier philosophical positions of the likes of Dewey (1916) and Peters (1981) , that all effective education is moral education, have been confirmed empirically in the updated neuroscientific evidence on which we are resting our case for the integral connections between creativity and critical reasoning.

The three mindsets to which Narvaez (2013) refers are safety, engagement and imagination, this latter associated most patently with the pre-frontal cortex of the brain. Despite its uniqueness, the operation of this dimension of the brain is not necessarily a positive thing; it can align itself with a security mindset and issue in 'vicious imagination'; this will most likely result if a person feels fear or insecurity, hence allowing imagination to impel negative thoughts and behaviours and, in turn, rigidifying brain activity. Bulkeley (2005) also notes the potential for fear resulting from the decentering that can accompany the experience of wonder. On the other hand, imagination can be allied with the engagement mindset and result in "communal imagination" (Narvaez, 2013, 2014), a confident and altruistic set of thoughts and behaviours that, among other things, enhances and enriches the growth and development of the brain.

There will be many factors in play that will determine which way imagination is directed, home life being one and the role of the teacher, school and school system being an important other dimension. Whichever social agency, the constant need is for a caring, trusting ambience. Bulkeley suggests of wonder that it requires "... a sufficient degree of safety and security ... so that the natural fear of being decentered does not smother the pleasurable creative response in seeking a new center" (p. 199). Narvaez (2014) pushes this point further to critique the whole ambience of contemporary Western society which, she argues, undermines communal imagination in its focus on threat and fear and she posits, instead, a return to the nurturing values of more primal cultures to better facilitate this aspect of human functioning (and in fact all of human flourishing). Both Narvaez and Bulkeley emphasize that it is the caring and trusting ambience, including the key people in any student's life that activates the kind of imagination and wonder that impels the emotional regulation that is essential to problem- solving, knowledge-seeking curiosity and critical reasoning. It is this kind of imagination and wonder that is associated with the capacity to engage in the deliberative and intuitive aspects of interaction and decision making. As such, education which places the value of the student at its centre—that is, a moral education—is best positioned to allow for the flourishing of creativity, imagination, and wonder.

Further to the important link between imagination, effective learning and moral motivation/ education, Narvaez (2013) proffers that communal imagination issues in a personality that "is able to move beyond immediate self- interest, to conceptualize alternative social systems, think impartially about moral problems, counteract harmful instincts or behave altruistically" (p. 331). Communal imagination is closely aligned with the capacity for empathy, which is itself "a primary force behind moral behaviour" (p. 327). In turn, it is the imagination mindset that gives rise to the higher-order thinking capacities needed "for problem solving and deliberative learning" (p. 327), especially when it comes to critical moral reasoning. In this regard, Eslinger et al. (2009) note that such functions are among the most complex undertaken by the brain because it must draw on more sophisticated neural sources when it encounters moral situations that include novel or ambiguous dimensions. Hence, a case is premised for the crucial conjunction between creativity, critical reasoning and moral education. Furthermore, we believe evidence exists that confirms that this conjunction is beyond mere speculative theorizing. We point now to findings emanating from the moral education venture titled the Australian Values Education Program.

CREATIVITY AND CRITICAL REASONING: EVIDENCE FROM THE AUSTRALIAN VALUES EDUCATION PROGRAM

As above, research insights and findings from the neurosciences illustrate the importance of education being holistic in its focus. The learner is a whole person, impelled by cognitive, emotional and social drives, not as separable features but in holistic connection with each other. Narvaez (2013) identifies the vital connections between the caring and supportive learning environment that, together with stimulating imagination and creativity, nurture the emotional regulation essential for sound (critical) reasoning powers to develop. Hence, is seen the crucial importance of pedagogy that places students at the centre of the learning enterprise by establishing encouraging, trusting and caring ambiences of learning and, furthermore, engaging in curricula that stimulate creativity, imagination and wonder. Such environments are those that research has shown to be the most likely to produce sustained and effective improvements in learning (Newmann et al., 1996 ; Bryk & Schneider, 2002 ; Rowe, 2004). There is now a vast store of evidence from research around moral education that the establishment of such ambiences of learning, together with explicit discourse about moral and values-oriented content in ways that draw on students' deeper learning and reflectivity, has power to transform their patterns of feelings, behaviour, resilience and academic diligence (cf. Hoffmann, 2000 ; Benninga et al., 2006 ; Lovat, 2010, 2011, 2013 ; Lovat & Toomey, 2009 ; Lovat et al., 2009, 2010a, 2010b, 2010c, 2011a, 2011b ; Berkowitz, 2011 ;Nucci & Narvaez, 2008). The principal research drawn on here as supporting evidence is taken from a range of the core projects that ran under the umbrella of the Australian Values Education Program.

The Australian Values Education Program was a federally funded venture that began with a pilot study in 2003 (DEST, 2003), followed by the development of a National Framework for Values Education in 2005 (DEST, 2005) and a range of attached research and practice projects from 2005 to 2009, the most crucial of which were the two stages of the Values Education Good Practice Schools Project (VEGPSP) (DEST, 2006 ; DEEWR, 2008) and the Project to Test and Measure the Impact of Values Education on Student Effects and School Ambience (Lovat et al., 2009).

Findings from VEGPSP (DEST, 2006 ; DEEWR, 2008) illustrated that a sound values education program can be a powerful ally in the development of best practice pedagogy, with positive effects being demonstrated across the range of measures, including in terms of academic development. Project reports identified greater reflectivity and strengthened intellectual engagement among students and settling into work more readily and calmly as routine effects of the ambience created when moral and values-oriented considerations were driving the pedagogical approach. Reports also spoke of the enhanced academic diligence that resulted. In most cases, reports of strengthened intellectual understanding pointed not only to improved learning performance but equally to enhanced moral behaviour. This was seen in clear demonstrations of a stronger sense of student responsibility over local, national and international issues, improved relationships of care and trust, greater student awareness of the need to be tolerant of others, to accept responsibility for their own actions and their ability to communicate, with students' sense of belonging, connectedness, resilience and sense of self all being enhanced.

In the Project to Test and Measure the Impact of Values Education on Student Effects and School Ambience (Lovat et al., 2009), a mixed methods approach in the form of a sequential explanatory design (Creswell et al., 2003) was taken in order to test all of the claims arising from VEGPSP concerning student effects. Quantitative data were collected over two time-periods and analysed, with qualitative data being collected during the second phase and analysed separately to help explain and elaborate on the quantitative results. The qualitative data helped to refine and explain the statistical results by incorporating more detailed information from the perspectives of the research participants. Student, staff and parent pre and post surveys were administered in order to obtain quantitative and qualitative data about the effects of the program on student behaviour, student engagement, and classroom and school ambience. The results of the analysis of the teacher surveys revealed statistically significant improvements on these three measures.

The qualitative data also supported the above findings, with many comments from both students and teachers indicating that improved interactions between students had led to more harmonious and productive learning environments in which students were demonstrating greater kindness to each other and taking more care and pride in their work. The teachers observed that giving students more control over routine tasks added to their sense of competence and this appeared to lead to more independent learning and increased intrinsic motivation, impelled by strengthened critical reasoning powers. The teachers reported that students were putting greater effort into their work and "striving for quality", "striving to achieve their best" and even "striving for perfection":

Thus, there was substantial quantitative and qualitative evidence suggesting that there were observable and measurable improvements in students' academic diligence, including increased attentiveness, a greater capacity to work independently as well as more cooperatively, greater care and effort being invested in schoolwork and students assuming more responsibility for their own learning (Lovat et al., 2009: p. 6).

Consistent with the focus of this paper is the emphasis in the evidence on creativity, imagination and wonder playing a vital role in achieving students' demonstrable improvement in the powers of critical reasoning. In speaking about the enhanced opportunities to deepen reflection and so strengthen learning overall, a teacher wrote:

I now see the need, when I am discussing ideas with children, not just to draw on the children's experience, but rather to extend their understanding of reality. I have a responsibility to provide time and experiences for children to wonder. We need to wonder ourselves. This will help us to understand and improve relationships with others (DEST, 2006: p. 34).

In another cluster of schools in which disadvantage was a particular issue, the researcher wrote of the rights of children to respect, care and quality education and, in this context, of the importance of nurturing "... creativity, wonderment (so they could) learn to participate actively in planning for their (own) future" (DEST, 2006: p. 208).

Meanwhile, one of the researcher "friends" of a cluster of schools utilizing a "philosophy in the classroom" tool for their values education objectives, identified these objectives as "... a school ethos where students have the capacity to exercise judgement and responsibility in matters of morality, ethics and social justice, and the capacity to make rational and informed decisions about their lives" (DEST, 2006: p. 109). In speaking to the nature of the approach being taken to achieve the objective, the researcher had the following to say about the added value to critical reasoning of creativity and wonder:

Philosophy is often described as a thinking skills programme or a course in critical and creative thinking. While it is true that philosophy for children does improve students' critical and creative thinking skills, calling it a "thinking skills" programme does not do it justice. It does much more as well. Through implementing Philosophy in our respective school sites we aim to build on the students' own wonder and curiosity about ideas that are important to them. The subject matter of Philosophy is those common, central and contestable concepts that underpin both our experience of human life and all academic disciplines ... The central pedagogical tool and guiding ideal of Philosophy is the community of inquiry. In the community of inquiry, students work together to generate and then answer their own questions about the philosophical issues contained in purpose written materials or a wide range of other resources. Thinking in the community of inquiry is critical, creative, collaborative and caring (DEST, 2006: p. 109).

CONCLUSION

The article has selected an array of prominent literature in an attempt to capture some of the vital evidence emanating from neuroscientific research about the central role of creativity, imagination and wonder in stimulating critical reasoning. It furthermore identifies moral education as a particularly powerful form of learning in impelling the kind of creativity, imagination and wonder necessary to the development of critical reasoning. Finally, the article points to empirical evidence from the Australian Values Education Program that appears to confirm the case being made.

Terence Lovat, Daniel Fleming As evidenced within the article, the authors are conscious of the contentiousness surrounding neuroscience's capacity to inform the practicalities of education, especially in the direct way that proposes biological findings can be ramified to teaching and learning as though no filter is required. We take note of the cautions provided by the likes of Bruer and Snook in this regard. In spite of such important cautions, we nonetheless come down on the side of the weight of scholarship that proffers important insights being generated by neuroscience for educational considerations, including concerned with the role of creativity, wonder and imagination in facilitating the kind of critical reasoning that impels educational achievement. As suggested within the article, we believe that the weight of the literature tends to the view that such insights from neuroscience are at the very least useful as metaphors for re-thinking many of the traditional assumptions about how learning can best proceed.

REFERENCES

1. ACARA (2013). General Capabilities in the Australian Curriculum. Canberra: Australian Curriculum Reporting and Assessment Authority. http://www.australiancurriculum.edu.au/GeneralCapabilities/General%20capabilities.pdf
2. Benninga, J., Berkowitz, M., Kuehn, P., & Smith, K. (2006). Character and Academics: What Good Schools Do. Phi Delta Kappan, 87, 448-452. http://dx.doi.org/10.1177/003172170608700610
3. Berkowitz, M. (2011). What Works in Values Education?

International Journal of Educational Research, 50, 153-158. http://dx.doi.org/10.1016/j.ijer.2011.07.003

4. Bielert, C., & Gallup, A. (2014). A Review of Darcia Narvaez, Kristin Valentino, Augustin Fuentes, James J. McKenna, and Peter Gray (Eds.), Ancestral Landscapes in Human Evolution: Culture, Childrearing and Social Wellbeing. New York: Oxford University Press. Evolutionary Psychology, 12, 1053-1055.
5. Bruer, J. (2006). Points of View: On the Implications of Neuroscience Research for Science Teaching and Learning: Are There Any? A Skeptical Theme and Variations: The Primacy of Psychology in the Science of Learning. CBE Life Sciences Education, 5, 104-110. http://dx.doi.org/10.1187/cbe.06-03-0153
6. Bruer, J. T. (1997). Education and the Brain: A Bridge Too Far. Educational Researcher, 26, 4-16. http://dx.doi.org/10.3102/0013189X026008004
7. Bruer, J. T. (1998). Let's Put Brain Science on the Back Burner. NASSP Bulletin, 82, 9-19. http://dx.doi.org/10.1177/019263659808259803
8. Bryk, A. (2012). Opening Remarks from the President at the US Professors of the Year Awards Program. Carnegie Foundation for the Advancement of Teaching. http://www.usprofessorsoftheyear.org/Winners/Previous_Natl_Winners/2012RemarksBryk.html
9. Bryk, A., & Schneider, B. (2002). Trust in Schools: A Core Resource for Improvement. New York: Russell Sage Foundation.
10. Bulkeley , K. (2005). The Wondering Brain: Thinking about Religion with and beyond Cognitive Neuroscience. New York: Routledge.
11. Clement, N. D., & Lovat, T. (2012). Neuroscience and Education: Issues and Challenges for Curriculum. Curriculum Inquiry, 42, 534-557. http://dx.doi.org/10.1111/j.1467-873X.2012.00602.x
12. Creswell, J. W., Plano Clark, V. L., Gutmann, M., & Hanson, W. (2003). Advanced Mixed Methods Research Designs. In A. Tashakkori, & C. Teddlie (Eds.), Handbook of Mixed Methods in Social and Behavioural Research (pp. 209-240). Thousand Oaks, CA: Sage.
13. DEEWR, Department of Education, Employment and Workplace Relations (2008). At the Heart of What We Do: Values Education at the Centre of Schooling. Report of the Values Education Good

Practice Schools Project—Stage 2. Melbourne: Curriculum Corporation. http://www.curriculum.edu.au/verve/_resources/VEGPSP-2_final_3.pdf
14. DEST, Department of Education Science and Training (2003). Values Education Study. Executive Summary Final Report, Melbourne: Curriculum Corporation. http://www.curriculum.edu.au/verve/_resources/VES_Final_Report14Nov.pdf
15. DEST, Department of Education Science and Training (2005). National Framework for Values Education in Australian Schools. Canberra: Australian Government Department of Education, Science and Training. http://www.curriculum.edu.au/verve/_resources/Framework_PDF_version_for_the_web.pdf
16. DEST, Department of Education Science and Training (2006). Implementing the National Framework for Values Education in Australian Schools. Report of the Values Education Good Practice Schools Project—Stage 1: Final Report, September 2006, Melbourne: Curriculum Corporation. http://www.curriculum.edu.au/verve/_resources/VEGPS1_FINAL_REPORT_081106.pdf
17. Dewey, J. (1916). Democracy and Education: An Introduction to the Philosophy of Education. New York: The Free Press.
18. Dietrich, A. (2004). The Cognitive Neuroscience of Creativity. Psychonomic Bulletin & Review, 11, 1011-1026. http://dx.doi.org/10.3758/BF03196731
19. Doidge, N. (2009). The Brain That Changes Itself: Stories of Personal Triumph from the Frontiers of Brain Science. Melbourne: Scribe Publications.
20. Dommett, E. J., Devonshire, I. M., Plateua, C. R., Westwell, M. S., & Greenfield, S. A. (2011). From Neuroscientific Theory to Classroom Practice. The Neuroscientist, 17, 282-288. http://dx.doi.org/10.1177/1073858409356111
21. Eslinger, P., Robinson-Long, M., Realmuto, J., Moll, J., de Oliveira-Souza, R., Tovar-Moll, F., Wang, J., & Yang, Q. X. (2009). Developmental Frontal Lobe Imaging in Moral Judgment: Arthur Benton's Ensuring Influence 60 Years Later. Journal of Clinical and Experimental Neuropsychology, 31, 158-169. http://dx.doi.org/10.1080/13803390802298064

22. Fleming, D., & Lovat, T. (2014). When Encounters between Religious Worldviews Are a Threat: Applying Triune Ethics Theory in a Religiously Diverse Landscape. Journal of Moral Education, 43, 377-393. http://dx.doi.org/10.1080/03057240.2014.945397
23. Fleming, D., & Lovat, T. (2015). Learning as Leaving Home: Fear, Empathy and Hospitality in the Theology Classroom. Teaching Theology and Religion, 18, in Press.
24. Friston, K., Ashburner, J., Kiebel, S., Nichols, T., & Penny, W. (2011). Statistical Parametric Mapping: The Analysis of Functional Brain Images. London: Academic Press.
25. Gortner, D. (2007). A Review of Kelly Bulkeley, The Wondering Brain: Thinking about Religion with and beyond Cognitive Neuroscience, New York: Routledge (2005). Anglican Theological Review, 89, 493-494.
26. Haidt, J. (2010). Moral Psychology Must Not Be Based on Faith and Hope: Commentary on Narvaez. Perspectives on Psychological Science, 5, 182-184. http://dx.doi.org/10.1177/1745691610362352
27. Hassabis, D., Kumaran, D., & Maguire, E. (2007). Using Imagination to Understand the Neural Basis of Episodic Memory. Journal of Neuroscience, 27, 14365-14374. http://dx.doi.org/10.1523/JNEUROSCI.4549-07.2007
28. Higgins-D'Alessandro, A. (2011). The Structures of a Liberal Education. Ethos, 10, 4-9.
29. Hille, K. (2011). Bringing Research into Educational Practice: Lessons Learned. Mind, Brain, and Education, 5, 63 70. http://dx.doi.org/10.1111/j.1751-228X.2011.01111.x
30. Hoffman, M. L. (2000). Empathy and Moral Development: Implications for Caring and Justice. Cambridge: Cambridge University Press. http://dx.doi.org/10.1017/cbo9780511805851
31. Immordino-Yang, M. H. (2011). Implications of Affective and Social Neuroscience for Educational Theory. Educational Philosophy and Theory, 43, 98-103. http://dx.doi.org/10.1111/j.1469-5812.2010.00713.x
32. Immordino-Yang, M. H., & Damasio, A. R. (2007). We Feel, Therefore We Learn: The Relevance of Affect and Social

Neuroscience to Education. Mind, Brain, and Education, 1, 3-10. http://dx.doi.org/10.1111/j.1751-228X.2007.00004.x

33. Immordino-Yang, M., & Faeth, M. (2010). Building Smart Students: A Neuroscience Perspective on the Role of Emotion and Skilled Intuition in Learning. In D. A. Sousa (Ed.), Mind, Brain, and Education: Neuroscience Implications for the Classroom (pp. 66-81). Bloomington, IN: Solution Tree.
34. Jensen , E. (2008). A Fresh Look at Brain-Based Education. Phi Delta Kappan, 89, 409-417. http://dx.doi.org/10.1177/003172170808900605
35. Lewis, P. (2010). The Emerging Comprehensive Psychology of Darcia Narvaez. Tradition & Discovery, 37, 9-18. http://dx.doi.org/10.5840/traddisc2010/201137332
36. Lovat, T. (2010). Synergies and Balance between Values Education and Quality Teaching. Educational Philosophy and Theory, 42, 489-500. http://dx.doi.org/10.1111/j.1469-5812.2008.00469.x
37. Lovat, T. (2011). Values Education and Holistic Learning: Updated Research Perspectives. International Journal of Educational Research, 50, 148-152. http://dx.doi.org/10.1016/j.ijer.2011.07.009
38. Lovat, T. (2013). Values Education Programs. In J. Hattie, & E. Anderman (Eds.), International Guide to Student Achievement (pp. 279-281). New York: Routledge.
39. Lovat, T., & Toomey, R. (Eds.) (2009). Values Education and Quality Teaching: The Double Helix Effect. Dordrecht: Springer. http://dx.doi.org/10.1007/978-1-4020-9962-5
40. Lovat, T., Clement, N., Dally, K., & Toomey, R. (2010b). Addressing Issues of Religious Difference through Values Education: An Islam Instance. Cambridge Journal of Education, 40, 213-227. http://dx.doi.org/10.1080/0305764X.2010.504599
41. Lovat, T., Clement, N., Dally, K., & Toomey, R. (2010c). Values Education as Holistic Development for All Sectors: Researching for Effective Pedagogy. Oxford Review of Education, 36, 1-17. http://dx.doi.org/10.1080/03054985.2010.501141
42. Lovat, T., Clement, N., Dally, K., & Toomey, R. (2011a). The Impact of Values Education on School Ambience and Academic Diligence. International Journal of Educational Research, 50, 166-171. http://

dx.doi.org/10.1016/j.ijer.2011.07.008

43. Lovat, T., Dally, K., Clement, N., & Toomey, R. (2011b). Values Pedagogy and Student Achievement: Contemporary Research Evidence. Dordrecht: Springer. http://dx.doi.org/10.1007/978-94-007-1563-9
44. Lovat, T., Toomey, R., & Clement, N. (Eds.) (2010a). International Research Handbook on Values Education and Student Wellbeing. Dordrecht: Springer. http://dx.doi.org/10.1007/978-90-481-8675-4
45. Lovat, T., Toomey, R., Dally, K., & Clement, N. (2009). Project to Test and Measure the Impact of Values Education on Student Effects and School Ambience. Report for the Australian Government Department of Education, Employment and Workplace Relations (DEEWR) by the University of Newcastle, Australia, Canberra: DEEWR. http://www.curriculum.edu.au/verve/_resources/Project_to_Test_and_Measure_the_Impact_of_Values_Education.pdf
46. MCEETYA (2008). Declaration on Educational Goals for Young Australians. Melbourne: Ministerial Council on Education, Employment, Training & Youth Affairs. www.curriculum.edu.au/verve/_resources/National_Declaration_on_the_Educational_Goals_for_Young_Australians.pdf
47. McGilchrist, I. (2009). The Master and His Emissary: The Divided Brain and the Making of the Western World. New Haven, CT: Yale University Press.
48. Meltzoff, A., Kuhl, P., Movellan, J., & Sejnowski, T. (2009). Foundations for a New Science of Learning. Science, 325, 284-288. http://dx.doi.org/10.1126/science.1175626
49. Narvaez, D. (2008). Triune Ethics: The Neurobiological Roots of Our Multiple Moralities. New Ideas in Psychology, 26, 95-119. http://dx.doi.org/10.1016/j.newideapsych.2007.07.008
50. Narvaez, D. (2010). The Emotional Foundations of High Moral Intelligence. New Directions for Child and Adolescent Development, 2010, 77-94. http://dx.doi.org/10.1002/cd.276
51. Narvaez, D. (2013). Neurobiology and Moral Mindset. In K. Heinrichs, F. Oser, & T. Lovat (Eds.), Handbook of Moral Motivation: Theories, Models, Applications (pp. 323-342). Rotterdam: Sense

Publishers. http://dx.doi.org/10.1007/978-94-6209-275-4_19

52. Narvaez, D. (2014). Neurobiology and the Development of Human Morality: Evolution, Culture, and Wisdom. New York: W. W. Norton & Company.
53. Narvaez, D., Panksepp, J., Schore, A. N., & Gleason, T. R. (2013). Evolution, Early Experience and Human Development: From Research to Practice and Policy. Oxford: Oxford University Press.
54. Narvaez, D., Valentino, K., Fuentes, A., McKenna, J., & Gray, P. (Eds.) (2014). Ancestral Landscapes in Human Evolution: Culture, Childrearing and Social Wellbeing. New York: Oxford University Press. http://dx.doi.org/10.1093/acprof:oso/9780199964253.001.0001
55. Newmann, F. (2013). Muzzling Minds All over the Globe While Trumpeting Higher Order Skills. http://www.invisibleserfscollar.com/tag/fred-newmann/
56. Newmann, F. M., & Associates (1996). Authentic Achievement: Restructuring Schools for Intellectual Quality. San Francisco, CA: Jossey-Bass Publishers.
57. Nucci , L., & Narvaez, D. (Eds.) (2008). Handbook of Moral and Character Education. New York: Routledge.
58. Peters, R. S. (1981). Moral Development and Moral Education. London: George Allen & Unwin.
59. Raichle, M. E. (2009). A Brief History of Human Brain Mapping. Trends in Neurosciences, 32, 118-126. http://dx.doi.org/10.1016/j.tins.2008.11.001
60. Rosiek, J., & Beghetto, R. (2009). Emotional Scaffolding: The Emotional and Imaginative Dimensions of Teaching and Learning. In P. Schutz, & M. Zembylas (Eds.), Advances in Teacher Emotion Research: The Impact on Teachers' Lives (pp. 175-194). Dordrecht: Springer. http://dx.doi.org/10.1007/978-1-4419-0564-2_9
61. Rowe, K. (2004). In Good Hands? The Importance of Teacher Quality. Educare News, 149, 4-14.
62. Sajjadi, S. M. (2008). Religious Education and Information Technology: Challenges and Problems. Teaching Theology and Religion, 11, 185-190. http://dx.doi.org/10.1111/j.1467-9647.2008.00469.x

63. Schrag, F. (2011). Does Neuroscience Matter for Education? Educational Theory, 61, 221-237. http://dx.doi.org/10.1111/j.1741-5446.2011.00401.x

64. Snook, I. (2012). Educational Neuroscience: A Plea for Radical Scepticism. Educational Philosophy and Theory, 44, 445- 449. http://dx.doi.org/10.1111/j.1469-5812.2011.00831.x

65. Srinivasan, N. (2007). Cognitive Neuroscience of Creativity: EEG Based Approaches. Science Direct, 42, 109-116. http://dx.doi.org/10.1016/j.ymeth.2006.12.008

66. Varma, S., McCandliss, B., & Schwartz, D. (2008). Scientific and Pragmatic Challenges for Bridging Education and Neuroscience. Educational Researcher, 37, 140-152. http://dx.doi.org/10.3102/0013189X08317687

67. Wegerif, R., McLaren, B. M., Chamrada, M., Scheuer, O., Mansour, N., Mikšátko, J., & Williams, M. (2010). Exploring Creative Thinking in Graphically Mediated Synchronous Dialogues. Computers & Education, 54, 613-621. http://dx.doi.org/10.1016/j.compedu.2009.10.015

68. Weisberg, D., Keil, F., Goodstein, G., Rawson, E., & Gray, J. (2008). The Seductive Allure of Neuroscience Explanations. Journal of Cognitive Neuroscience, 20, 470-477. http://dx.doi.org/10.1162/jocn.2008.20040

SECTION III: PATTERN RECOGNITION IN NEURO AND MEDICAL APPLICATIONS

CHAPTER 8

MACROSCOPIC NEURAL OSCILLATION DURING SKILLED REACHING MOVEMENTS IN HUMANS

Hong Gi Yeom,[1] June Sic Kim,[2] and Chun Kee Chung[1,2,3]

[1]Interdisciplinary Program in Neuroscience, Seoul National University, Seoul 151-742, Republic of Korea

[2]Department of Brain and Cognitive Sciences, College of Natural Sciences, Seoul National University, Seoul 151-742, Republic of Korea

[3]Department of Neurosurgery, Seoul National University Hospital, Seoul 110-744, Republic of Korea

ABSTRACT

The neural mechanism of skilled movements, such as reaching, has been considered to differ from that of rhythmic movement such as locomotion. It is generally thought that skilled movements are consciously controlled by the brain, while rhythmic movements are usually controlled autonomously by the spinal cord and brain stem. However, several studies in recent decades have suggested that neural networks in the spinal cord may also be involved in the generation of skilled movements. Moreover, a recent study revealed that neural activities in the motor cortex exhibit rhythmic oscillations corresponding to movement frequency during reaching movements as rhythmic movements. However, whether the oscillations are generated in the spinal cord or the cortical circuit in the motor cortex causes the oscillations is unclear. If the spinal cord is involved in the skilled movements, then similar rhythmic oscillations with time delays should be found in macroscopic neural activity. We measured whole-brain MEG signals during reaching. The MEG signals were analyzed using a dynamical analysis method. We found that rhythmic oscillations with time delays occur in all subjects during reaching movements. The results suggest that the corticospinal system is involved in the generation and control of the skilled movements as rhythmic movements.

INTRODUCTION

The neural mechanism of skilled movements has been considered to differ from that of rhythmic movement [1]. Skilled movements, such as reaching and grasping, are nonperiodic and are consciously controlled by the brain, while rhythmic movements such as locomotion are repetitive and stereotypical. Although rhythmic movements can be controlled voluntarily, these movements are usually controlled autonomously by the spinal cord and brain stem. A central pattern generator (CPG) in the spinal cord produces periodic oscillatory patterns [1, 2]. The CPG has been considered to be associated with the control of rhythmic movement [3–5].

However, it has been suggested that not only the cortical circuit but also the neural networks in the spinal cord may be involved in skilled movements [6–12]. Moreover, Rokni and Sompolinsky demonstrated that various natural movements can be generated by the linear summation of simple oscillatory components [13]. When considering Fourier theory, it is reasonable to presume that all of the complicated signals can be approximated by the linear summation of sine and cosine signals [14]. The suggestion of generating the various movements from simple components also corresponds to the perspective of dynamic systems, suggesting that most neural activity in the motor cortex will be internal processes that drive desired movements [15]. A recent study reported an important phenomenon. The study, based on neural dynamical analysis, demonstrated that rhythmic oscillations corresponding to the movement frequency also occur during skilled reaching movements [16] as rhythmic movements [16–19]. This implies that diverse skilled movements can be generated via CPG, similar to the neural mechanism of rhythmic movements.

However, in the previous study, a very small motor area in a monkey was measured at a microscopic level. Therefore, whether the rhythmic oscillations are generated in CPG or the cortical circuit in the motor cortex causes the oscillations is unclear. Moreover, the occurrence of the rhythmic oscillations corresponding to the movement frequency has not been confirmed in humans.

Various pathways connect the cortex and spinal cord. The direct corticomotoneuronal (CM) pathway connects the motor cortex to spinal motoneurons. Indirect pathways might connect other sensorimotor cortices, such as the premotor (PM), supplementary (SMA), cingulate (CMA), and primary somatosensory (S1) areas, to the spinal cord [20]. Therefore, if the rhythmic oscillations occur from the spinal cord and are delivered to the broad motor-related cortex, similar rhythmic oscillations with time delays should be found in macroscopic neural activity (Figures 3(a) and 3(b)). Here, we examine whether similar rhythmic oscillations with time delays are exhibited in macroscopic neural activity during reaching movements in humans. To investigate neural activity, we measured whole-brain MEG signals during reaching movements. We analyze the MEG signals using an analysis method, j principle component analysis (jPCA), where j implies an imaginary part in a complex conjugate. The method reveals the dynamical characteristics of the neural activity [16]. If there are similar oscillatory patterns with time delay, the projections of the oscillations onto the jPC planes will be rotated. Therefore, we can easily investigate whether similar rhythmic oscillations occur or not by examining the projections. Moreover, the results will indicate whether the rhythmic oscillations occur from spinal cords or not.

MATERIALS AND METHODS

Experiment and Data Acquisition

Nine healthy subjects (age: 19–37 years; five males and four females) participated in the experiment. All subjects were right-handed (Edinburgh Handedness Inventory scores were above 80). A 306-channel whole-head MEG system (VectorView™, Elekta Neuromag Oy, Helsinki, Finland) was used to measure neural activity during reaching movements (Figure 1(a)). The MEG system has 306 sensors grouped in triplets consisting of 2 planar gradiometers and 1 magnetometer distributed at 102 locations. To record arm position, a three-axis accelerometer (KXM52, Kionix, NY, USA) was attached to the index finger of the right hand using the Velcro band. The accelerometer signals were recorded simultaneously with the MEG signals. The sampling frequency of the MEG and the accelerometer signals was 600.615 Hz. The experiment was approved by the Institutional Review Board of the Seoul National University Hospital (IRB number 1105-095-363). Subjects were instructed to perform a center-out reaching task according to stereographic images on a screen. To minimize movement artifacts, a cushion was placed under the subject's elbow during the experiment.

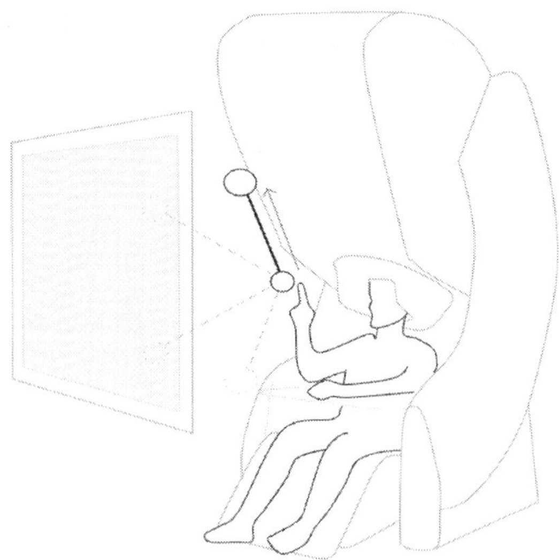

(a)

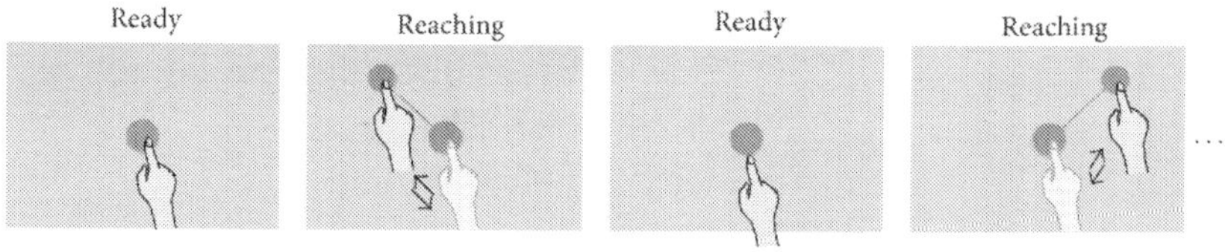

(b)

Figure 1: Experiment paradigm. (a) Illustration of the virtual visual stimuli and MEG acquisition system. MEG signals and hand positions were recorded simultaneously during reaching movements. Subjects were instructed to perform a center-out reaching task according to stereographic images on a screen. The target sphere appeared randomly on one of the four corners (upper-left, upper-right, bottom-left, and bottom-right). The red arrow illustrates the example of the reaching movements when the target is shown at the upper-left. (b) Drawings showing the sequence of visual stimuli and instructed behaviors. At the beginning of the experiments, a sphere was shown on the center of the screen. After 4 s, the target was presented for 1 s on one of the four corners. During this time, the subjects were instructed to shift their index finger from the center to the target and return to the center according to the connecting line by moving their right arm as fast as possible. This sequence was repeated during the experiments.

At the beginning of the experiments, a sphere was shown on the center of the screen for 4 s and a target sphere with a stick connecting it to the center sphere was presented on one corner for 1 s. The target sphere appeared randomly on one of the four corners (upper-left, upper-right, bottom-left, and bottom-right). During this time, the subjects were instructed to shift their index finger from the center to the target and return to the center according to the connecting line by moving their right arm as fast as possible (Figure 1(b)). This sequence was repeated during the experiments. For each subject, 60 trials were measured for each direction. The distances from the center to the target were ~20 cm. Because we used stereographic images to represent the target, the distance measurement was not accurate. The distances from the center to the target were identical in all subjects. Because the directions of the target were randomly presented, we did not consider the variation of the intertrial interval for habituation effects. Although the reaction times were slightly different (see Section 3), there is no strange trial. Further details are also described elsewhere [21, 22].

Signal Preprocessing

The MEG equipment measures the changes of a magnetic field. Therefore, it is influenced by external signals, such as line noise, and biological artifacts, such as cardiac and muscle activity. To reduce the noise in the MEG signals, the spatiotemporal signal space separation (tSSS) method was applied [23]. The tSSS separates external interference signals of the brain by spatiotemporal methods and eliminates the interference. All data processing was performed using MATLAB 2008b (Mathworks, Natick, MA, USA). We used 204 gradiometer signals among the 306 channels for data analysis because the characteristics of the gradiometer and magnetometer sensors differ and the signal-to-noise ratio (SNR) of the gradiometers is better than that of the magnetometers [24]. The MEG signals were band-pass filtered between 0.5 and 8 Hz. The filtering band was determined by time-frequency analysis based on our previous study [21]. The MEG signals of the frequency band represent the characteristics of reaching movements [21]. To minimize artifacts, independent component analysis (ICA) was applied, which is implemented in EEGLAB [25]. Artifacts such as EOG were removed by eliminating the artifact components. The signals were segmented from −1 to 2 s after the cue onset. After the segmentation, the MEG signals were averaged by trials. The averaged MEG signals were downsampled to 50 Hz. jPCA was applied to the preprocessed signals. The process of the analysis is explained in the next section and in Figure 2.

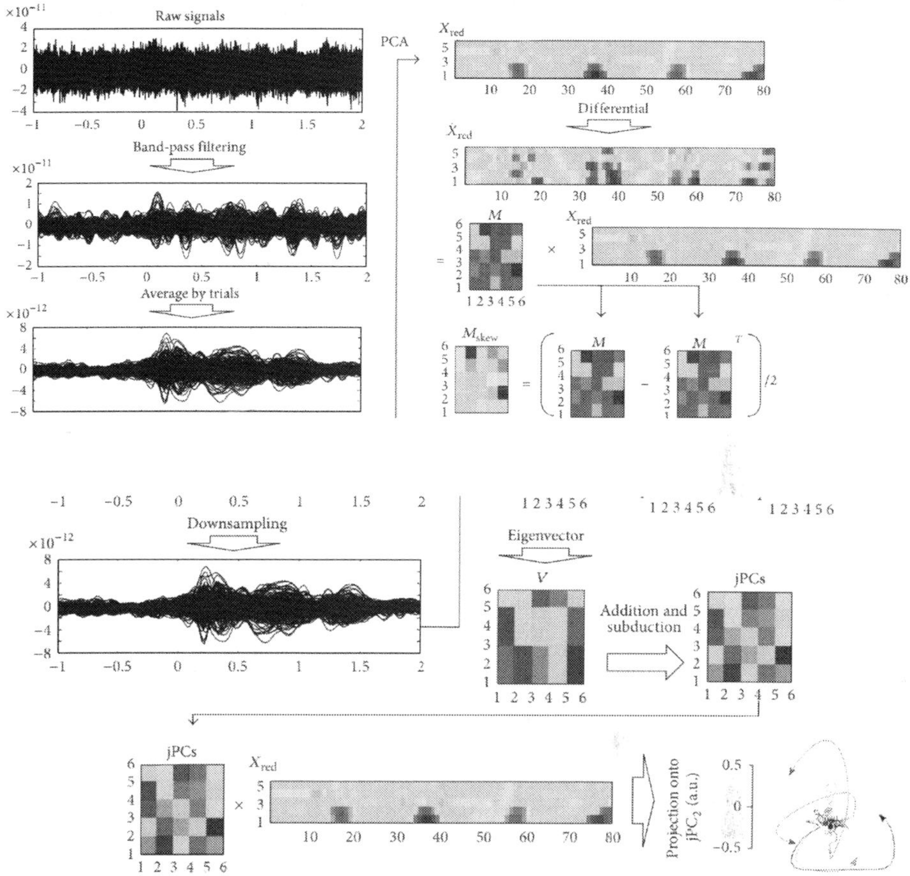

Figure 2: Signal process of jPCA analysis. The segmented MEG signals from 1-2 s before the cue onset were band-pass filtered between 0.5 and 8 Hz. After filtering, the MEG signals were averaged by trials. The averaged MEG signals were downsampled at 20 ms. Matrices M and M_{skew} were calculated from downsampled signals. The eigenvectors of M_{skew} produced jPCs. Projections on the jPC plane were illustrated by multiplying jPCs and X_{red}.

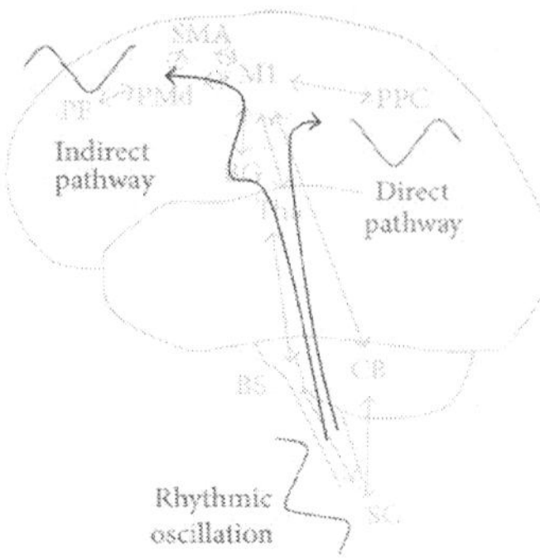
SMA
M1
PPC
PF
PMd
Indirect pathway
Direct pathway
BG
CB
BS
Rhythmic oscillation
SC

(a)

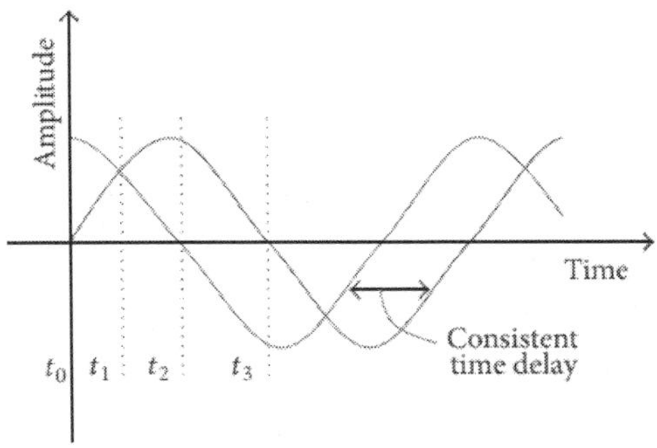
Amplitude
Time
Consistent time delay
t_0
t_1
t_2
t_3

(b)

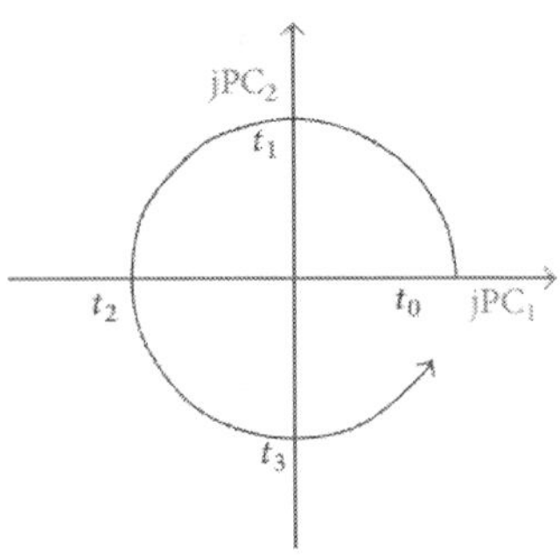
jPC_2
jPC_1
t_0
t_1
t_2
t_3

(c)

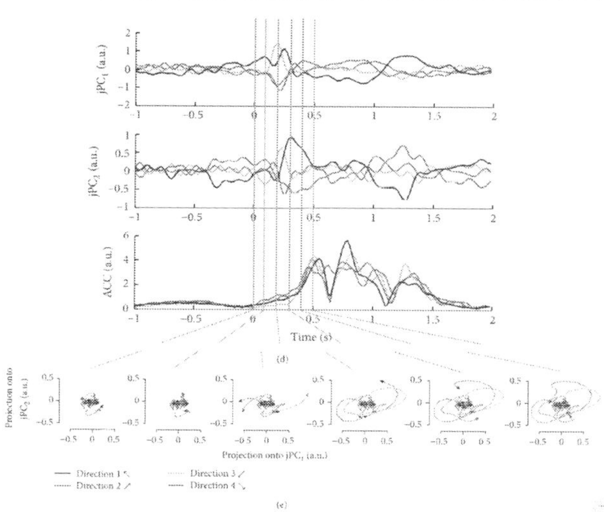

Figure 3: Illustration of a simple rhythmic oscillation model to draw a circle on the jPC planes and results of jPCA analysis. (a) Delivery of a rhythmic oscillation to the different area through direct and indirect pathways with different time delays. (b) Oscillations of the different areas with similar patterns and a consistent time delay. (c) Rotation of the oscillations on the jPC planes. (d) jPCs and root-mean squares (RMSs) of accelerometer signals with time. The different-colored signals represent the movements in different directions. 0 ms indicates the stimulus onset time. (e) Projections on the jPC planes are shown at 0, 100, 200, 300, 400, and 500 ms. The projections were rotated prior to movement onset. The results suggest that the oscillations are related not only to movement execution but also to movement preparation.

Dynamical Analysis jPCA

Neural networks in the brain consist of billions of neurons. Although the activities of some neurons in the motor cortex will reflect movement parameters, most neural activities will be an internal process to generate the motor commands [15]. Therefore, the signals are difficult to analyze because the patterns of neural activity are various and complex. To investigate the neural process, a dynamical analysis method, jPCA, was proposed [16]. jPCA describes the dynamical relationship between current and subsequent neural activities on a low-dimensional plane. It represents the dynamic relationship with rotation according to the time flow. If there are consistent phase-differences between neural activities, the projections on the jPC planes will be rotated; however, if there is no consistent change, meaningless small movements will be shown. Moreover, if the sign of the phase-difference changes, the projections will be rotated in the opposite direction. This reveals the change in the dynamic relationship between the current and subsequent neural activities. jPCA is an intuitive method of analyzing the dynamic characteristics of neural activities. jPCA finds informative planes and projects the neural data onto the planes. The projected neural data represents the rotational structure in the data.

The relationship between the current and next neural activities can be expressed as follows:

$$\dot{X} = MX \qquad (1)$$

where X is a matrix of size $n \times ct$ describing the neural activities. The preprocessed MEG signals as described in Section 2.2 were applied to jPCA as X. n is the number of MEG channels ($n = 204$). c is the number of conditions (c=4), and t is the number of time points ($t = 151$). X is a derivative of X and M represents the relationship between the neural activity and its derivative.

jPCA applies the traditional principal component analysis (PCA) to reduce the dimensionality of X from $n \times ct$ to $k \times ct$ (k=6). The reduced neural activity will be expressed as Xred. Equation (1) can be represented as follows:

$$\dot{X}_{\text{red}} = MX_{\text{red}}. \qquad (2)$$

M can be calculated by linear regression. M is a combination of symmetric transformation and skew transformation. M can be divided as follows:

$$M = M_{\text{symm}} + M_{\text{skew}} \qquad (3)$$

where M_{symm} and M_{skew} are defined as $M_{\text{symm}} = (M+M_T)/2$ and $M_{\text{skew}} = (M - M_T)/2$

Equation (2) can be expressed as follows:

$$\dot{X}_{red} = M_{skew} X_{red}. \tag{4}$$

Because M_{skew} has imaginary eigenvalues, it captures the rotational dynamics of neural activity. To express complex eigenvectors V_1 and V_2 of M_{skew} on a real plane, the jPC can be defined as $jPC_1 = V_1 + V_2$ and $jPC_2 = j(V_1 - V_2)$. The projection onto the jPC can be calculated as $X_{jPCA} = (jPC_1; jPC_2) \times X_{red}$. Further details are provided in [16]. Our results show the projections on one on each jPC (Figure 3(d)) or on the two-dimensional space of jPC_1 and jPC_2 (Figure 4).

RESULTS

Figure 3 shows a simple rhythmic oscillation model that draws a circle on jPC planes and results from jPCA. To draw a circle on jPC planes as in Figure 3(c), the rhythmic oscillations of jPC_1 and jPC_2 should have a similar pattern. Moreover, the time difference between oscillations should be consistent as in Figure 3(b). Figure 3(d) illustrates jPCs and root-mean square (RMS) of the accelerometer signals over time. Projections on jPC planes are also shown at 0, 100, 200, 300, 400, and 500 ms in Figure 3(e).The different colors of the signals represent the movements of different directions. 0 ms indicates the stimulus onset time. The average movement onset was 316 ± 58 ms (mean ± standard deviation) from stimulus onset. In contrast, the rhythmic oscillations began at 133±69 ms. We compared the onset times between the movements and the rhythmic oscillations using a paired t-test. The onsets differed significantly ($p < 0.001$) between the movements and the rhythmic oscillations. After the presentation of visual stimuli, the rhythmic oscillations of jPCs occurred before arm movements, as shown in Figure 3(d). Projections on jPC planes show clearer results. Before movement onset, the projections of the neural oscillations were rotated. Figure 4 shows the projections on jPC planes for each subject from −100 to 300 ms after presentation of visual stimuli. The duration corresponds to the time taken for movement preparation. The projection of the oscillation rotated in the same direction (counterclockwise) for all conditions (reaching different directions) for all subjects. To rotate in the same direction for different conditions, the relation between jPC1 and jPC2 should be invariant irrespective of the movement direction. Table 1 shows a summary of jPCA. The second row describes the data variance captured by jPC plane. The mean of the data variance was 0.235 ± 0.059. The third row is fit quality provided by M_{skew}, in the first two jPCs. The mean of the fit quality from two jPCs was 0.235 ± 0.075.

DISCUSSION

In this study, we demonstrated that the neural mechanisms of rhythmic movements and skilled movements are similar. We showed that the corticospinal system is involved in the generation of skilled movements by means of a dynamic analysis of macroscopic neural data. Although our result does not concur with previous knowledge, it supports the suggestion made in recent studies that the spinal cord mediates skilled movements [6–12].

Neural Mechanism of Skilled Movements

It is unclear whether the rhythmic oscillations are generated in the spinal cord or in the motor cortex. We hypothesize that if the rhythmic oscillations are derived by the corticospinal system, oscillations with similar pattern and consistent time delay should be found in macroscopic neural activity (Figures 3(a) and 3(b)). To investigate this hypothesis, we measured and analyzed MEG signals during reaching movements in humans. The results showed that projections on the jPC planes rotated in all subjects (Figure 4). This means that the major components of the neural activity have a similar pattern and consistent time delay. Therefore, it implies the possibility that skilled movements are generated by the corticospinal system. Moreover, the projection of the oscillation rotated in the same direction for all conditions. This suggests that the same neural dynamics are involved in reaching movements, irrespective of reaching direction. Our perspective corresponds to the suggestion of a common intrinsic structure controlling the reaching movements [16, 26]. When we consider that the cortex also contributes to controlling rhythmic movements [7, 27], the neural mechanisms of rhythmic movements and skilled movements are similar. Despite numerous studies of a CPG in animals, almost all studies in humans used indirect results [28]. We also did not measure spinal cord activity directly.Therefore, we should be cautious when interpreting the results. Nevertheless, our results provide evidence that the corticospinal system is involved in skilled movement, such as rhythmic movement.

Neural Oscillations Involved in Movement Generations

Iteration of descending motor commands and ascending sensory feedback could be represented by rhythmic patterns. Therefore, the rhythmic oscillations could be considered products generated by sensory feedback, such as kinematic parameter or visual feedback [28–32]. However, our results show that the rhythmic oscillation occurred prior to movement onset (Figures 3(d), 3(e), and 4). Because sensory feedback could not generate the oscillations before the movements, it implies that there is another mechanism of generating the rhythmic oscillations. It has been suggested that neural networks in the spinal cord may also be involved in skilled movements. We hypothesized that if the rhythmic oscillations are made by CPG and delivered to the broad motor-related area through direct and indirect pathways, similar patterns of time delay will be observed at a macroscopic level. We verified our hypothesis using the dynamical analysis method.Therefore, our results suggest that the spinal cord could be involved in the movement generation. When a subject is aware of a target position in a motor planning task, the reaction time required to reach the target after the go cue was about 240 ms [33]. Because the visual processing requires less than 150 ms [34], and the executionrelated time will be about 90 ms. In our experiment, the movement onset time (316 ms) may include the time for visual processing, motor planning, and execution. Therefore, the difference (76 ms) between 316 ms and 240 ms is related to motor planning. Thus, 166 ms (76 + 90 ms) may be required for movement planning and execution. The rhythmic oscillation occurred 183 ms (316 − 133 ms) prior to movement onset, suggesting that the rhythmic oscillations are related to movement generation.

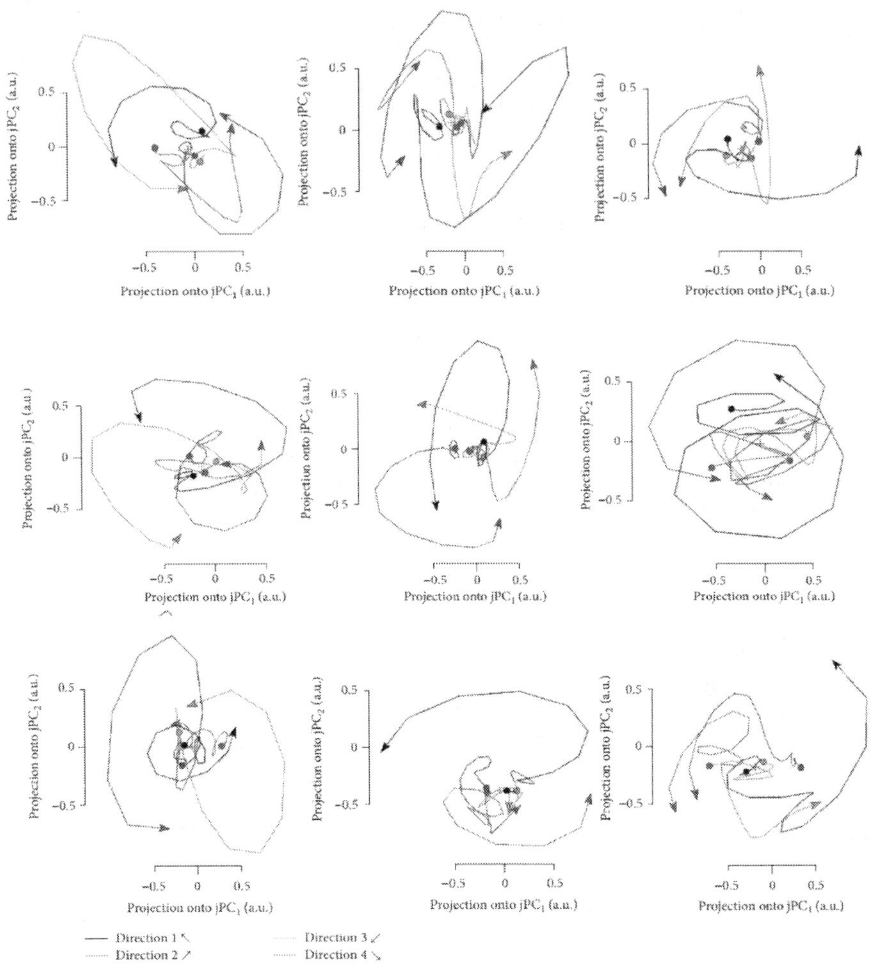

Figure 4: Projections on the jPC planes for all subjects from −100 to 300 ms after presentation of visual stimuli. Each subfigure illustrates the projections from each subject. The different-colored signals represent the movements in different directions. The projection of the oscillation rotated to the same direction (counterclockwise) for all conditions (reaching of different direction) for all subjects. This suggests that the same neural dynamics are involved in reaching movements irrespective of reaching direction.

Table 1: Summary of jPCA. Captured variance, data variance captured by the jPC plane; fit by 2 jPCs, fit quality provided by M_{skew} in the first two jPCs.

	Sub 1	Sub 2	Sub 3	Sub 4	Sub 5	Sub 6	Sub 7	Sub 8	Sub 9	Mean ± standard deviation
Captured variance	0.226	0.382	0.180	0.212	0.190	0.231	0.242	0.243	0.212	0.235 ± 0.059
Fit by 2 jPCs	0.164	0.226	0.157	0.168	0.245	0.250	0.237	0.401	0.271	0.235 ± 0.075

CONCLUSIONS

We showed that neural oscillations occur at a macroscopic level in humans during skilled movements. It seems that a common intrinsic structure generates the oscillations, irrespective of movement direction. The intrinsic structure is involved in not only movement execution but also movement generation. The neural oscillations could be generated in the spinal cord and the oscillations might influence movements by means of corticospinal interaction. This implies that the neural mechanism of skilled movements might be similar to that of rhythmic movements.

Competing Interests

The authors declare that they have no competing interests.

Acknowledgments

This study was supported by Grant (NRCTR-EX15002) of the Translational Research Center for Rehabilitation Robots, Korea National Rehabilitation Center, Ministry of Health & Welfare, Korea.

REFERENCES

1. E. R. Kandel, Principles of Neural Science, McGraw-Hill, New York, NY, USA, 5th edition, 2013.
2. K. G. Pearson, "Common principles of motor control in vertebrates and invertebrates," Annual Review of Neuroscience, vol. 16, pp. 265–297, 1993.
3. M. MacKay-Lyons, "Central pattern generation of locomotion: a review of the evidence," Physical Therapy, vol. 82, no. 1, pp. 69–83, 2002.
4. E. Marder and D. Bucher, "Central pattern generators and the control of rhythmic movements," Current Biology, vol. 11, no. 23,

pp. R986–R996, 2001.

5. E. Marder and R. L. Calabrese, "Principles of rhythmic motor pattern generation," Physiological Reviews, vol. 76, no. 3, pp. 687–717, 1996.
6. B. Alstermark and T. Isa, "Circuits for skilled reaching and grasping," Annual Review of Neuroscience, vol. 35, pp. 559–578, 2012.
7. T. Drew, S. Prentice, and B. Schepens, "Cortical and brainstem control of locomotion," Progress in Brain Research, vol. 143, pp. 251–261, 2004.
8. A. P. Georgopoulos and S. Grillner, "Visuomotor coordination in reaching and locomotion," Science, vol. 245, no. 4923, pp. 1209–1210, 1989.
9. E. Azim, J. Jiang, B. Alstermark, and T. M. Jessell, "Skilled reaching relies on a V2a propriospinal internal copy circuit," Nature, vol. 508, no. 7496, pp. 357–363, 2014. ·
10. B. Alstermark, L. G. Pettersson, Y. Nishimura et al., "Motor command for precision grip in the macaque monkey can be mediated by spinal interneurons," Journal of Neurophysiology, vol. 106, no. 1, pp. 122–126, 2011.
11. B. Alstermark, T. Isa, L.-G. Pettersson, and S. Sasaki, "The C3-C4 propriospinal system in the cat and monkey: a spinal pre-motoneuronal centre for voluntary motor control," Acta Physiologica, vol. 189, no. 2, pp. 123–140, 2007.
12. T. Isa, Y. Ohki, B. Alstermark, L.-G. Pettersson, and S. Sasaki, "Direct and indirect cortico-motoneuronal pathways and control of hand/arm movements," Physiology, vol. 22, no. 2, pp. 145–153, 2007.
13. U. Rokni and H. Sompolinsky, "How the brain generates movement," Neural Computation, vol. 24, no. 2, pp. 289–331, 2012. View at Zentralblatt MATH · View at MathSciNet ·
14. R. N. Bracewell, The Fourier Transform and Its Applications, McGraw-Hill, Boston, Mass, USA, 3rd edition, 2000.
15. K. V. Shenoy, M. Sahani, and M. M. Churchland, "Cortical control of arm movements: a dynamical systems perspective," Annual Review of Neuroscience, vol. 36, pp. 337–359, 2013.

16. M. M. Churchland, J. P. Cunningham, M. T. Kaufman et al., "Neural population dynamics during reaching," Nature, vol. 486, no. 7405, pp. 51–56, 2012.

17. W. B. Kristan Jr. and R. L. Calabrese, "Rhythmic swimming activity in neurones of the isolated nerve cord of the leech," Journal of Experimental Biology, vol. 65, no. 3, pp. 643–668, 1976.

18. K. L. Briggman and W. B. Kristan Jr., "Imaging dedicated and multifunctional neural circuits generating distinct behaviors," Journal of Neuroscience, vol. 26, no. 42, pp. 10925–10933, 2006.·

19. K. L. Briggman, H. D. I. Abarbanel, and W. B. Kristan Jr., "Optical imaging of neuronal populations during decision-making," Science, vol. 307, no. 5711, pp. 896–901, 2005.

20. T. Isa, M. Kinoshita, and Y. Nishimura, "Role of direct vs. indirect pathways from the motor cortex to spinal motoneurons in the control of hand dexterity," Frontiers in Neurology, vol. 4, article 191, 2013. ·

21. H. G. Yeom, J. S. Kim, and C. K. Chung, "Estimation of the velocity and trajectory of three-dimensional reaching movements from non-invasive magnetoencephalography signals," Journal of Neural Engineering, vol. 10, no. 2, Article ID 026006, 2013.

22. H. G. Yeom, J. S. Kim, and C. K. Chung, "Common neural mechanism for reaching movements," in Proceedings of the 4th International Winter Conference on Brain-Computer Interface (BCI ‹16), pp. 1–3, Gangwon Province, Republic of Korea, Feburary 2016.

23. S. Taulu and J. Simola, "Spatiotemporal signal space separation method for rejecting nearby interference in MEG measurements," Physics in Medicine and Biology, vol. 51, no. 7, pp. 1759–1768, 2006.

24. M. Hämäläinen, R. Hari, R. J. Ilmoniemi, J. Knuutila, and O. V. Lounasmaa, "Magnetoencephalography—theory, instrumentation, and applications to noninvasive studies of the working human brain," Reviews of Modern Physics, vol. 65, no. 2, pp. 413–497, 1993.

25. A. Delorme and S. Makeig, "EEGLAB: an open source toolbox for analysis of single-trial EEG dynamics including independent

component analysis," Journal of Neuroscience Methods, vol. 134, no. 1, pp. 9–21, 2004.

26. T. M. Hall, F. de Carvalho, and A. Jackson, "A common structure underlies low-frequency cortical dynamics in movement, sleep, and sedation," Neuron, vol. 83, no. 5, pp. 1185–1199, 2014.
27. T. Drew, J.-E. Andujar, K. Lajoie, and S. Yakovenko, "Cortical mechanisms involved in visuomotor coordination during precision walking," Brain Research Reviews, vol. 57, no. 1, pp. 199–211, 2008.
28. M. R. Dimitrijevic, Y. Gerasimenko, and M. M. Pinter, "Evidence for a spinal central pattern generator in humans," Annals of the New York Academy of Sciences, vol. 860, pp. 360–376, 1998.
29. K. J. Craik, "Theory of the human operator in control systems; the operator as an engineering system," The British Journal of Psychology, vol. 38, no. 2, pp. 56–61, 1947.
30. R. C. Miall, D. J. Weir, and J. F. Stein, "Manual tracking of visual targets by trained monkeys," Behavioural Brain Research, vol. 20, no. 2, pp. 185–201, 1986.
31. R. C. Miall, D. J. Weir, and J. F. Stein, "Intermittency in human manual tracking tasks," Journal of Motor Behavior, vol. 25, no. 1, pp. 53–63, 1993.
32. R. C. Miall and J. K. Jackson, "Adaptation to visual feedback delays in manual tracking: evidence against the Smith Predictor model of human visually guided action," Experimental Brain Research, vol. 172, no. 1, pp. 77–84, 2006.
33. C. A. Chestek, A. P. Batista, G. Santhanam et al., "Single-neuron stability during repeated reaching in macaque premotor cortex," The Journal of Neuroscience, vol. 27, no. 40, pp. 10742–10750, 2007.
34. S. Thorpe, D. Fize, and C. Marlot, "Speed of processing in the human visual system," Nature, vol. 381, no. 6582, pp. 520–522, 1996.

CHAPTER 9

BRAINK FOR STRUCTURAL IMAGE PROCESSING: CREATING ELECTRICAL MODELS OF THE HUMAN HEAD

Kai Li,[1] Xenophon Papademetris,[1,2] and Don M. Tucker[1]

[1]Electrical Geodesics, Inc., 500 E. 4th Avenue, Eugene, OR 97401, USA

[2]Yale University, New Haven, CT 06520, USA

ABSTRACT

BrainK is a set of automated procedures for characterizing the tissues of the human head from MRI, CT, and photogrammetry images. The tissue segmentation and cortical surface extraction support the primary goal of modeling the propagation of electrical currents through head tissues with a finite difference model (FDM) or finite element model (FEM) created from the BrainK geometries. The electrical head model is necessary for accurate source localization of dense array electroencephalographic (dEEG) measures from head surface electrodes. It is also necessary for accurate targeting of cerebral structures with transcranial current injection from those surface electrodes. BrainK must achieve five major tasks: image segmentation, registration of the MRI, CT, and sensor photogrammetry images, cortical surface reconstruction, dipole tessellation of the cortical surface, and Talairach transformation. We describe the approach to each task, and we compare the accuracies for the key tasks of tissue segmentation and cortical surface extraction in relation to existing research tools (FreeSurfer, FSL, SPM, and BrainVisa). BrainK achieves good accuracy with minimal or no user intervention, it deals well with poor quality MR images and tissue abnormalities, and it provides improved computational efficiency over existing research packages.

INTRODUCTION

With the simple recording of the electroencephalogram (EEG), the brain's electrical activity can be measured with millisecond temporal resolution at the head surface. Dense array EEG (dEEG) systems now allow up to 256 channels to be applied quickly with full coverage of the head, assessing the fields from the basal as well as superior cortical surface [1, 2]. The cortex, with its laminar neural organization and with locally synchronous activity stemming from its columnar organization, is the primary generator of the far fields measured by head surface EEG [3]. Cortical sources can be modeled as point dipoles, and their contribution to surface activity can be reconstructed through electrical source analysis. The ambiguity of the inverse estimation in electrical source analysis can be minimized if the precise locations and orientations of the cortical sources are well specified. To a first approximation, the source dipoles can be assumed to be oriented perpendicular to the cortical surface, consistent with the orientation of the pyramidal neurons and cortical columns.

To model cortical sources with these properties, accurate cortical surface extraction is a key challenge. Furthermore, the volume conduction of the electrical potentials, from the cortex to the head surface, must be specified through characterizing the conductivity of each tissue compartment. The skull is the primary resistive medium in the head, and it must be modeled, preferably with bone density values from CT. Because the electrical boundary effects of the volume conduction are affected by discontinuities in current paths, for example, caused by holes in the skull (optical canals and foramen magnum), spherical shell or boundary element models provide only approximate electrical propagation from cortex to the surface, and more detailed (FDM or FEM) volumetric models are needed. Finally, the position of the electrodes must be specified accurately, for example, with geodesic photogrammetry [4].

By describing the positions of electrodes and cortical surface targets, an accurate electrical head model also supports dense array approaches to transcranial current injection. The electrical head model can be validated with bounded Electrical Impedance Tomography (bEIT), in which current injection and potential recovery are analyzed, within the bounds of the tissue geometry from MRI, to test whether the electrical conductivity of head tissues estimated by the model is accurate in predicting the recovered potentials [5, 6]. With an accurate electrical head model, it is possible to modulate brain activity noninvasively, using dense array transcranial Direct Current Stimulation (tDCS), or transcranial Alternating Current Stimulation (tACS), in which patterns of source and sink electrodes are computationally optimized to target specific cerebral sites [7, 8].

In the present report, we review the methods implemented in BrainK. In addition, we report validation studies for accuracy and efficiency of both tissue segmentation and cortical surface extraction. Klauschen et al. [9] evaluated automatic tools for head tissue segmentation from MR images. These included FSL [10], FreeSurfer [11, 12], and SPM [13]. Although BrainK is designed as an automated tool, visualization and editing capability is provided to allow adaptation to unique image properties, such as the presence of lesions or tissue anomalies.

OVERVIEW OF BRAINK

Figure 1 shows the graphical user interface for BrainK, including the multiple steps of image processing. Although maximal accuracy for EEG source localization or transcranial neuromodulation requires the full complement of MRI, CT, and sensor photogrammetry, it is important to optimize the results possible with the imaging data available for a particular subject or patient. Figure 2 shows the architecture of BrainK that incorporates specific workflows designed to adapt to the available data for the person. Although it is possible to assume standard sensor positions for BrainK (when sensor placement is referenced to skull fiducials, as with the Geodesic Sensor Net), a more accurate EEG source localization workflow begins with sensor positions from photogrammetry, such as what is provided by the Geodesic Photogrammetry System (GPS). If no imaging data is available for the person, an Atlas head model, constructed from a database of MRIs and CTs for the appropriate ages [14], is then warped with nonlinear registration to fit the sensor positions (and thus the person's head shape), producing a conformal Atlas head model. If only a structural MRI (typically a volumetric T1) is available for the person, the sensor positions are registered to the MRI, a database skull (from CT) is registered to the MRI, and tissue segmentation and cortical surface extraction are conducted to create theindividual head model with Atlas skull. The skull compartment includes either the original CT Hounsfield units for estimating bone density (and thus conductivity) voxel by voxel or a single skull segment, depending on the FDM computational model that will use BrainK's head model output. If both the CT and MRI are available for the person, the MRI is registered to the CT, which has more accurate dimensions than the MRI, to create theindividual head model with individual CT skull.

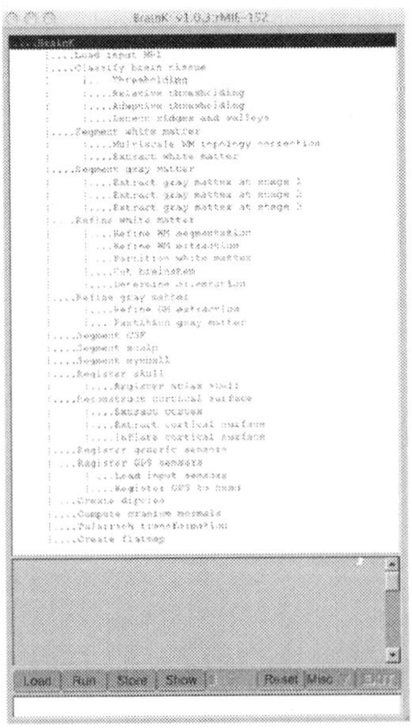

Figure 1: BrainK GUI.

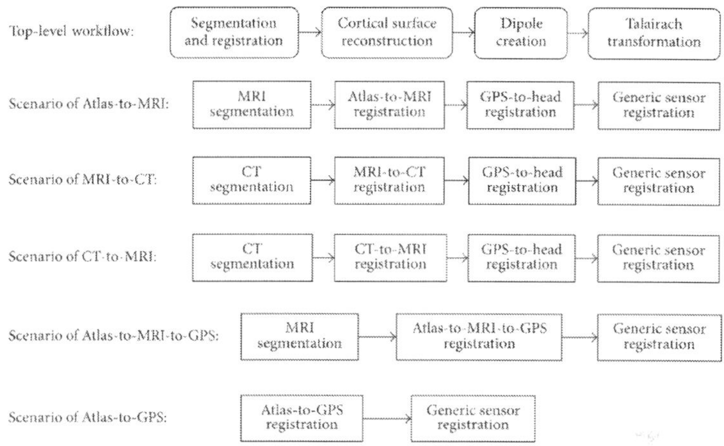

Figure 2: BrainK architecture.

For all workflows when the T1 MRI is available for the person, BrainK segmentation identifies the white matter (WM) and the gray matter (GM). It partitions these into two hemispheres and it differentiates cerebellum from cerebrum. In addition, an entire head mask and the two eyeballs are separated as well in the segmentation component. The eyeballs may be important for the electrical head model because of the large far fields generated by their cornea-retinal potentials (that must be separated from the brain signal in the EEG). For all workflow scenarios, a generic spherical sensor cloud is warped onto the head contour of the subject. For the scenarios of Atlas-to-MRI, MRI-to-CT, and CT-to-MRI, an additional GPS-to-head registration procedure is conducted to register the specific GPS sensor cloud, from one individual Geodesic Sensor Net application, onto the head contour. For the scenarios of Atlas-to-MRI-to-GPS and Atlas-to-GPS, the individual GPS data has already played a role in the skull registration and has been aligned with the head contour. In all scenarios, the skull registration is conducted, such that the resulting head segmentation includes the following tissue types: WM, GM, CSF, bone, flesh, and eyeball, in which the WM and the GM are further partitioned into two cerebral hemispheres and the cerebellum.

METHODS IN BRAINK

Image Segmentation

The MRI segmentation implements a cascade of automatic segmentation procedures in order to identify and separate the head tissues required for electrical head modeling, including the scalp, skull, cerebrospinal fluid (CSF), brain (gray and white matter), and the eyeballs. Finally, cortical surface extraction is performed to allow characterization of the normal of the cortical surface (allowing dipoles to be fit perpendicular to the cortex). The cascade mainly consists of two types of procedures: voxel classification on a region of interest (ROI) and the extraction of certain anatomical structures. The classification procedure aims at labeling all voxels in the ROI into different tissue types. The extraction procedure performs morphological operations on the voxel classification so that a certain anatomical structure, such as the WM, the GM, or the scalp, is separated from erroneous (false positive) segmentation results.

Given the MRI data, the segmentation cascade as shown in Figure 3 first takes the entire MR image space as the first ROI, R_1, and classifies all the voxels into two types: foreground and background. The following foreground extraction procedure does simple morphological operations to further improve the foreground segmentation resulting in the second ROI, R_2, in which all voxels are classified into three tissue types: WM, GM, and CSF, as shown in Figure 4(b). The major operation performed in this procedure is BrainK's novel relative thresholding (RT) technique (Section 3.1.1) [15]. Given the voxel classification on R_2, the WM extraction procedure then produces an initial WM volume W_1, which is used as the basis to identify the initial GM volume G_1 by the GM extraction procedure. Next, the union of G_1 and W_1 is taken as the third ROI, R_3, in which the WM/GM classification is refined with a different RT scheme. The new WM volume is then processed by the WM extraction procedure and a WM partition procedure, ending up with the second WM volume W_2.

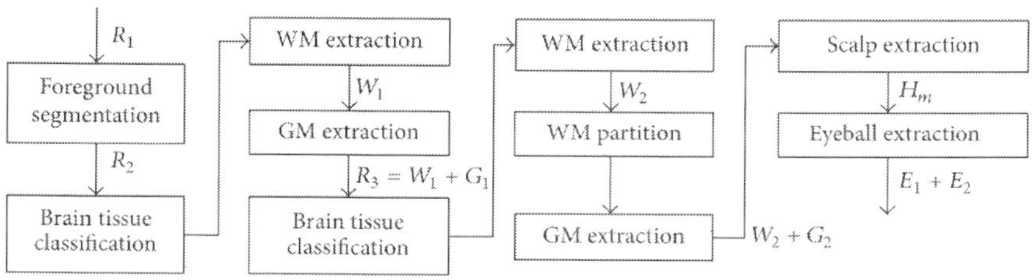

Figure 3: MRI segmentation workflow.

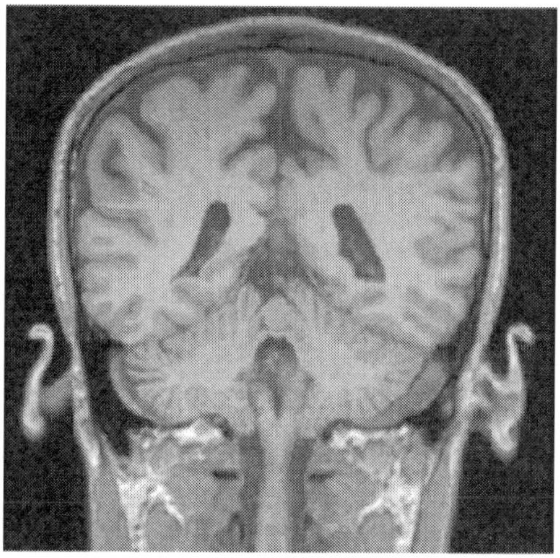

(a)

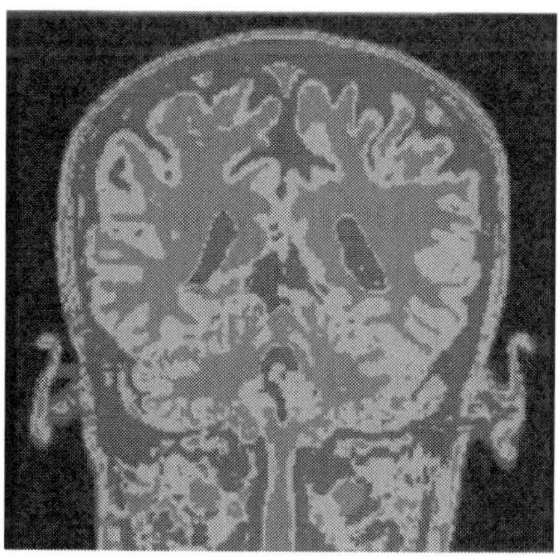

(b)

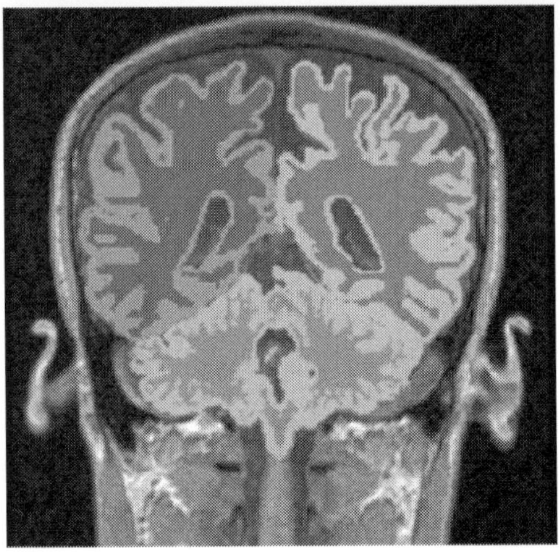

(c)

Figure 4: MRI segmentation. (a) MR image. (b) RT result on the foreground. (c) MRI segmentation result.

This WM partition procedure not only separates the two cerebral hemispheres from each other but also makes an optimal cut between the cerebral WM and the cerebellar WM with the well-known maximum flow algorithm [16].The result is specification of the following tissue types in W_2: two types of cerebral WM for the two hemispheres, respectively, and the cerebellar WM.W_2 then forms the basis for extracting the new GM volume G_2, which includes the two types of cerebral GM, respectively, for the two hemispheres and the cerebellar GM, as shown in Figure 4(c). The brain tissue segmentation (W_2 plus G_2) is then taken as a reference data for the scalp segmentation resulting in the head mask Hm. The scalp segmentation then conducts a binary thresholding procedure on R_1

To achieve the above segmentation steps, BrainK implements a novel morphological image analysis (SMIA) technique (Section 3.1.3) and a cell complex based morphometric image analysis (CCMIA) method (Section 3.1.4). These play key roles in the WM extraction, topology correction, GM extraction, and scalp extraction. Together with the relative thresholding RT method, the SMIA and CCMIA steps form the core techniques used in BrainK and will be described in the following subsections in detail. The eyeball extraction from the head mask is a special procedure using a priori knowledge about the eyeballs. Although segmentation of the skull from the MRI is an important challenge, the characterization of bone properties is poor in the typical T1 MRI sequence compared to that provided by CT. The Hounsfield units provided by the CT image (measuring X-ray attenuation) are directly proportional to bone density and thus provide important information on the relative tissue properties of trabecular skull. For example, the CT Hounsfield units correlate $r = 0.83$ with water content in trabecular skull [17]. The CT image segmentation procedure first performs the simple yet robust thresholding [18] so that the entire image is classified into three voxel types: bone, flesh, and background. The original Hounsfield units are then retained as bone density and conductivity estimates. Morphology operations follow to smooth the initial segmentation. Finally, the maximum flow algorithm is applied to separate the brain volume from the rest of the soft tissue so that the cut between them is minimized. CT segmentation results in a voxel classification as well as a brain volume wrapped up by the cranium and eventually there are four tissue types in the segmentation: bone, flesh, brain, and background.

Relative Thresholding (RT)

A well-known and prominent image artifact that challenges MR image segmentation is the intensity inhomogeneity (IIH) due to spatial distortions in the radio frequency (RF) gain in the RF coil [19]. The presence of IIH leads to a shading effect over the image and to significant overlap between histograms of different tissues. The result is that intensity based methods, such as thresholding and clustering, are typically unreliable in brain tissue segmentation. IIH correction can be performed prior to image segmentation, but the procedure may eliminate not only the image artifacts but also image signals that provide important information. When the correction is performed together with the segmentation, additional degrees of freedom are introduced into the problem formulation, making the optimization procedure more susceptible to local optima. In this paper, we present a new segmentation method referred to as relative thresholding (RT), which uses two global relative thresholds to compare with the local intensity contrasts, thereby bringing global and local information to segment the ROI into WM, GM, and CSF. The nature of RT makes it robust against the influence of IIT without introducing an explicit IIH correction. Like traditional thresholding methods, RT enables exhaustive searching without possibilities of being trapped into local optima. It also incorporates various a priori knowledge such that it is robust to intersubject variability and to the convolution of cortical structures. Finally, RT conducts brain tissue segmentation without the need of a prior step of brain volume extraction. In spite of intersubject variability, complexity of cortical structures, and variability on MR imaging sequences, we can make the following a priori structural, geometrical, and radiological observations: (1) Skull, CSF, GM, and WM form a layered structure from outside to inside; (2) the average intensities of skull, CSF, GM, and WM in local regions are in ascending order in T1-weighted MRI; (3) the cortex thickness is nearly uniform and is a very small value compared to the size of brains; in addition to the structural modeling described above, we also formulate an image model by incorporating a multiplicative low-frequency bias field b_i and an additive noise field ρ_i as follows:

$$y_i = b_i y_i' + \rho_i, \quad (1)$$

where yi is the observed intensity at the ith voxel and y'_i is the unobservable true intensity without the influences of IIH and noise.

The structural modeling and the image modeling enable two procedures, both of which compare the intensity of a subject voxel x_i with that of a reference voxel x_j on its gradient path, which is a set of ordered voxels emanating from x_i and following the gradient direction at each voxel in the path. The first procedure, GM-WM differentiation, scans all voxels in the ROI, which are initialized as WM, and labels them as GM if there is such a reference voxel x_j that its distance from x_i is less than a distance threshold d_{gw} and the ratio between their

smoothed intensities z_i/z_j is less than a relative threshold t_{gw}; the second procedure, CSF-GM differentiation, scans each GM voxel x_i and relabels it as CSF if there is such a WM voxel xk on its gradient path whose distance from x_i is less than another distance threshold d_{cw} ($d_{cw} > d_{gw}$) and the ratio between their smoothed intensities z_i/z_k is less than another relative threshold tcw ($t_{cw} < t_{gw}$).

In order to find the optimal relative thresholds, BrainK exhaustively tries each pair among the Cartesian product of the candidate GM/WM relative threshold set and the candidate CSF/WM relative threshold set and chooses the pair that minimizes an objective function $h = w_w h_w + w_g h_g + w_c h_c$. In this equation, h_w is a metric measuring the homogeneity of the WM object and formulated as the sum of smoothed intensity differences of any pair of adjacent voxels labeled as WM. hg and h_c are formulated in the same spirit for GM and CSF, respectively. w_w, w_g, and w_c are three weighting coefficients. The second RT scheme takes the optimal relative thresholds and differs from the first scheme on the question of how to determine the reference intensity compared to the intensity of the subject voxel. The ROI R_3 is first reset and then an initial set W^0 of WM voxels are determined by thresholding R_3 with a traditional threshold that maximizes the objective function $f = f_e - f_r$, where f_e is the sum of intensity differences of all m pairs of adjacent voxels labeled as WM and non-WM and f_r is the sum of differences of the m pairs of adjacent voxels labeled as WM and with the greatest intensity differences. A minimal cardinality of W^0 is set to improve the robustness. The remaining region of R_3 is then segmented as a procedure of iteratively dilating W^0 by using the optimal GM/WM relative threshold to compare the subject voxel's intensity to a reference intensity computed by considering the WM voxels near the subject voxel.

GM and WM Extraction

The GM and WM extraction procedures are responsible for extracting the GM and the WM structure from the raw voxel classification in the ROI. We describe GM extraction first, assuming that WM extraction is already done. The GM extraction utilizes the following a priori knowledge: (1) the thickness of cortex is nearly uniform; (2) GM wraps around WM such that two tissues form a layered structure; (3) the average gray level of GM is lower than that of WM in any local region. Given the extracted WM volume $W2$, the GM volume is first created by a gradient flow process, in which any voxel originally labeled as GM in the ROI is taken as the true GM if it can reach any WM voxels by following the gradient path emanating from itself within a given distance threshold. Each step in the gradient path is determined by looking at the gradient direction for each voxel. This initial GM volume is then further processed by traditional morphological operations to improve the segmentation, ending up with the GM volume *G*2. WM extraction is a more challenging process than GM extraction because it is responsible for extraction of WM itself, and it is also responsible for the brain volume extraction. The following a priori knowledge forms the basis of WM extraction: the WM is highly connected, but it has no topological defects. Topological defects or false positives present narrow bottlenecks. The extraction mainly consists of two steps: (1) WM localization, in which the center of one of the hemispheres is determined; (2) separation of WM from false positives. The localization of the WM center is essentially finding the WM voxel with the highest connectivity, which means that some quantitative measurement of connectivity should be taken for all WM voxels. Since topological defects, such as a tunnel inside the WM, can greatly influence the geometric measurement of the connectivity, the WM should be first topology-corrected before the connectivity measurement. Because the cortex surface (ribbon) must be defined in relation to the correct topology of the WM, WM topology correction is also essential for valid cortical surface reconstruction. BrainK introduces a skeletonization based morphological image analysis (SMIA) method and a cell complex based morphometric image analysis (CCMIA) framework. The following sections describe how these methods play central roles in WM extraction through achieving accurate connectivity measurement and topology correction.

Skeletonization Based Morphological Image Analysis (SMIA)

BrainK's SMIA framework consists of a surface skeletonization procedure and a curve skeletonization procedure, both of which are based on its extensive topological point classification. Given a binary image, we say its foreground *F* has a handle whenever there is a closed path in *F* that cannot be deformed through connected deformations in *F* to a single point. A handle in *F* is referred to as a tunnel in its complement $\vec{F}$•A point in *F* is simple if it can be added to or removed from *F* without changing the topology of both *F* and $\vec{F}$•A simple point [20] is a central concept in digital topology and is the basis of our definition of what we call a thick-simple point. A point *P* is thick-simple with respect to *F* if it is simple with respect to *F* and the removal of *P* and any of its neighbors from *F* does not increase the number of tunnels and number of connected components in $\vec{F}$•A point is thinsimpleif it is simple but not thick-simple and can be classified as thick-surface and thick-curve points. Thick-simple points can also be further classified as surface-edge, curve-end, and other point types.

There are two steps for BrainK's surface skeletonization based on the above point classification: thick-surface skeletonization and thin-surface skeletonization.The former results in a discrete surface with thickness of at most two voxels, and the latter results in a final thin surface skeleton of onevoxel thickness. The thick-surface skeletonization iteratively removes boundary points of the object that are thick-simple at the moment of removal. The thin-surface skeletonization iteratively removes boundary points of the thick-surface skeleton that are thick-surface points, thick-surface-edge points, or thick curve-end points at the moment of removal. The curve skeletonization procedure iteratively removes the boundary points of the object that are thick-curve points or thick-simple, but not curve-end points, at the moment of removal. As a side product, BrainK's surface skeletonization procedure gives each skeleton point a depth metric as the distance from the point to the boundary, whereas the curve skeleton procedure gives each skeleton point a wideness metric. If the curve skeletonization procedure is conducted at a certain scale, then the points that become the thin-curve points can be checked if they are on a handle of the object. Removal of such handle points then corrects the topology defects by cutting the handle. BrainK supports a multiscale topology correction [21] method on the WM object so that the WM extraction is made more robust, and then the topology-corrected cortex can be generated based on the WM.

Cell Complex Based Morphometric Image Analysis (CCMIA)

CCMIA is motivated by the goal of representing the true connectivity of an object's skeleton considering the convolution of such objects as the cerebral WM, where neither the depth nor the wideness of the structure is good enough for this need. It is essentially a series of transformations on a space called the cell complex [22]. A cell complex is a topological space composed of points, segments, polygons, polyhedrons, and the generation to polytope in any dimension. Given a set of voxels X in the 3D binary image, we can construct a 3-dimensional cell complex, that is, 3-complex, by creating a point for each voxel, an edge for every two connected points, a triangle for every three interconnected edges, and a tetrahedron for every four interconnected triangles. CCMIA proceeds as transforming the 3-complex to a 2- complex with points, segments, and polygons only and then the 2-complex to a 1-complex with points and segments only. The 3-2 transformation iteratively removes those boundary tetrahedrons by dropping one face f while a connectivity metric and a depth metric are accumulated on each f' of the remaining three faces for the removed tetrahedron as $\text{depth}(f') = \text{depth}(f) + d(f, f')$, while d represents the distance between the centers of the two faces. In a similar spirit, the 2-complex can be transformed to 1-complex and the remaining edges can be set with two accumulated metrics as $\text{wideness}(e') = \max(\text{widness}(e'), \text{wideness}(e) + d(e, e'))$ and $\text{connectivity}(e') = \text{connectivity}(e') + \text{connectivity}(e) + d(e, e')$.

After the raw WM object is topologically corrected, its surface skeleton is processed by CCMIA ending up with the 1-complex with points and edges only. The point incident to the edge of the greatest connectivity is identified as the WM center and a wideness threshold is used to break the bottleneck between the truc WM and the false positives. Finally, the WM surface skeleton is dilated to restore the volumetric object.

Registration

Given the varying data available, flexible and accurate registration is an essential BrainK function. Whereas the MR image accurately differentiates soft tissues, the CT image accurately represents bone. When individual CT or even MRI is not available, BrainK provides registration of the sensor photogrammetry data to a digital head tissue Atlas of the appropriate age and gender. BrainK integrates both soft tissues and bones from multimodal images and from existing digital Atlases by various image registration techniques. Image registration is a problem of finding optimal geometric transformations between images so that each point of one image can be mapped to a corresponding point of another image. There are generally rigid transformations and non rigid ones. Rigid transformation involves only translation and rotation. Affine transformation is a typical nonrigid transformation and allows scaling and shearing. Another form of non rigid transformation is an elastic transformation, allowing local deformation based on models from elasticity theory. BrainK implements rigid and affine transformations, as well as a landmark-based elastic transformation using thin plate spline theory [23]. For different registration purposes, either one or more of them is used so that the simple transformation serves as initial state for the more sophisticated transformation. Both the rigid transformation and the affine transformation can be represented by a simple 3×4 matrix. The key issue in applying such transformations is to optimize the similarity measure between the source and the target image, which in BrainK is represented in terms of landmarks. For the thin-plate spline transformation, the displacement field at any point is computed by interpolation from the vectors defined by a set of source landmarks and a set of target landmarks. It can be seen that landmark extraction plays a central role in all three transformations. BrainK is optimized for extracting landmarks based on the preceding image segmentation and on a priori anatomical knowledge. All registration procedures in BrainK undergo three steps: (1) landmarks extraction, (2) transformation coefficients determination, and (3) transformation execution. Note that three transformations all use landmarks but in different ways. The landmarks in rigid and affine transformation are used for constructing the similarity measurement, while there are source landmarks and target landmarks in thin-plate spine transformations and they are mapped to build the displacement field.

The thin-plate spline theory is based on an analogy to the approximate shape of thin metal plates deflected by normal forces at discrete points. Given n source landmarks $Pi = (x_i, y_i, z_i)$ in the source image and the corresponding target landmarks $P'_i = (x'_i, y'_i, z'_i)$ in the target image, the thin-plate spline transformation maps any point $P = (x, y, z)$ in the source image to the point $P' = (x', y', z')$ in the target image as follows:

$$x' = a_0 + a_1 x + a_2 y + a_3 z + \sum_{i=1} d_i r_i^2 \ln r_i^2,$$

$$y' = b_0 + b_1 x + b_2 y + b_3 z + \sum_{i=1}^{n} e_i r_i^2 \ln r_i^2,$$

$$z' = c_0 + c_1 x + c_2 y + c_3 z + \sum_{i=1}^{n} f_i r_i^2 \ln r_i^2, \qquad (2)$$

where $r^2_i = (x - x_i)^2 + (y - y_i)^2 + (z - z_i)^2$. The coefficients in the above equations together form a $(n + 4) \times 3$ matrix W, which can be obtained by solving the equation $LW = M$ that forces the matching of P_i and P'_i, where M is formed by organizing the coordinates of $P'_i = (x'_i, y'_i, z'i)$ and L is formed by composing the coordinates $P_i = (x_i, y_i, z_i)$ and the distance information $r^2_i \ln r^2_i$.

When the individual CT as well as the MRI data is available for the person, the registration component in the MRI-to-CT scenario first segments the CT into the following tissue types: bone, brain, flesh, and background. The MRI segmentation is then aligned with the CT segmentation so that the brain tissues and the eyeballs recognized in the MRI data are properly transformed and put into the brain region defined by the CT data. The result of this transformation is that the entire head segmentation (WM, GM, CSF, bone, eyeball, and flesh) is now registered with the CT volume, where the CSF is redefined as the "brain" region recognized in the CT segmentation, but now subtracting the WM and GM that were inserted from the MRI segmentation.

When both MRI and CT are available, the user could also choose the CT-to-MRI registration so that the bone recognized in the CT is transformed and put into the MRI segmentation wrapping around the brain and the CSF is inserted between the brain and the cranium. However, the MRI-to-CT registration is preferred when the CT covers the entire head of the subject because MRI imaging introduces more geometry distortion than CT. However, when the CT only covers the cranium region and the MRI data includes the face and the jaw, the user could consider the CT-to-MRI registration in order to have adequate face and jaw data for the electrical head model.

When only the MRI data is available for the given subject, BrainK supports the skull Atlas-to-MRI registration, in which the bone tissue from an Atlas dataset is warped into the MRI space and the CSF is also taken as those regions between the cranium and the brain. Any region within the head mask that is not occupied by the GM, the WM, the CSF, the eyeballs, or the bone is set to flesh.

When only the MRI and the individual GPS data are available and if the MRI is badly warped (dimensionally distorted), the user could choose the MRI-to-GPS registration, so that the head region is warped to match the GPS sensor cloud, which has good metric properties of the actual head shape. The result is transformed segmentation data, which can be further processed by the Atlas-to-MRI registration described above. Thus, the skull registration can be adapted to fit the available image data for the individual. When only the individual GPS data is available for a given subject, an Atlas dataset with all required tissue types can be warped to match the GPS sensor cloud with the so-called Atlas-to GPS registration, producing an individual conformal Atlas that has the appropriate shape of the volume conductor for electrical head modeling.

MRI/CT Registration

After MRI segmentation, we already have the MRI brain volume that is composed of WM and GM, but not CSF. The CT brain volume occupies spaces of WM, GM, and CSF. By applying morphological closing at a proper scale on the union of the MRI WM and GM, we obtain the MRI brain mask B_m, which has some inner CSF space filled and the brain contour smoothed but still misses some CSF voxels mainly wrapping around the superior contour of the brain. MRI segmentation also produces the MRI head mask H_m and the two eyeball masks E_1 and E_2. Correspondingly, we have the CT brain mask B_c and the CT head mask H_c. These data form the input set used to determine the transformation coefficients of the MRI/CT registration.

MRI/CT registration undergoes a rigid transformation followed by an affine transformation. Different sets of landmarks are extracted for the two transformations. The rigid transformation aims at only providing a good initial state for the affine transformation and its landmarks are simply those bordering voxels around $_{Bm.}$ The similarity measure to optimize the rigid transformation is the sum of squares of the distances from all landmarks to the contour of Bc. The optimal coefficients are found by multiscale exhaustive searching with acceptable computational efficiency and good global optimization performance.

The landmarks for the affine transformation are more sophisticated and consist of two parts: the inferior landmarks and the superior landmarks. The inferior landmarks are from the border voxels of the inferior part of B_m while the superior landmarks are from the superior contour of H_m. The inferior and the superior landmarks are partitioned so that they "visually" wrap up B_m but do not overlap. The motivation of using two sets of landmarks is to align MRI with CT in terms of both the brain matching and the scalp matching. However, the brain volume in the CT may contain significantly more CSF than that in the MRI at the superior part of the volume and we make the observation that their contours have reliable matching only at the inferior part of the brain. The landmark extraction thus essentially involves the partitioning of the cephalic space into the inferior part and the superior part. BrainK takes the following three landmarks as reference points to partition the cephalic space: the center of the brain mask B_m and the centers of the two eyeball masks E_1 and E_2. All three points can be automatically determined.

The similarity measure for the affine transformation is formulated as the sum of two terms. One is the sum of squares of the distances from all inferior landmarks to the contour of Bc and the other is the sum of squares of the distances from all superior landmarks to the contour of Hc. A gradient-descent optimization method is used to obtain the optimal affine coefficients. When all the optimal coefficients of both the rigid and the affine transformation are obtained, the MRI/CT registration can be performed from one direction while the other can be achieved by simply inverting the coefficient matrix. The result of MRI-to-CT registration is shown in Figure 5.

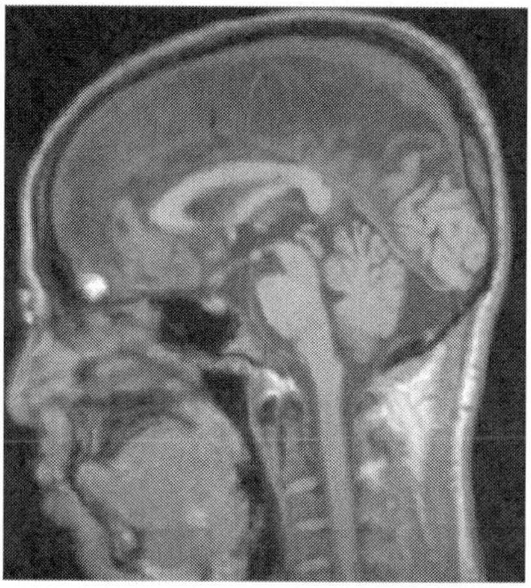

(a)

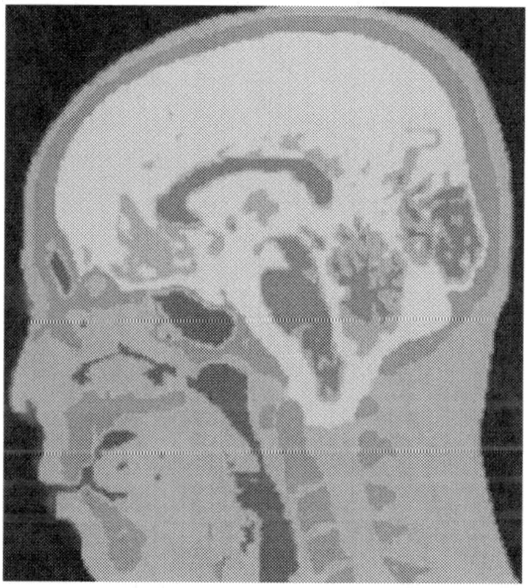

(b)

Figure 5: MRI-to-CT registration. (a) Transformed MRI. (b) Fusion of MRI and CT segmentation.

Skull Atlas-to-MRI Registration

Since the Atlas and the individual subject can have significant geometric differences, the Atlas-to-MRI registration undergoes an elastic thin plate spline transformation following the rigid and the affine transformations. The rigid transformation provides a good initial state for the affine transformation, which in turn provides a good initial state for the elastic local transformation. The result of the Atlas-to-MRI registration is illustrated in Figure 6.

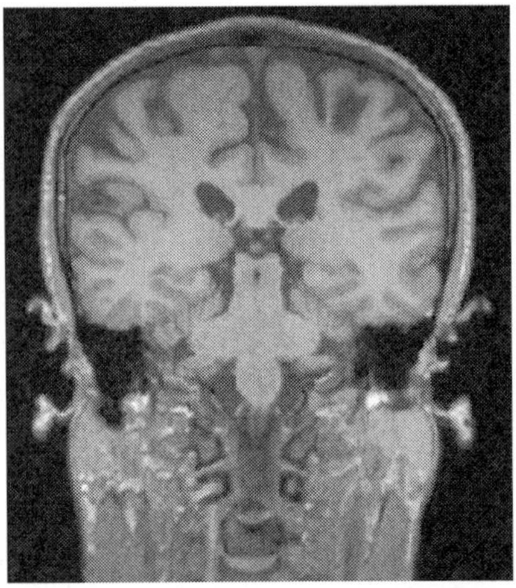

(a)

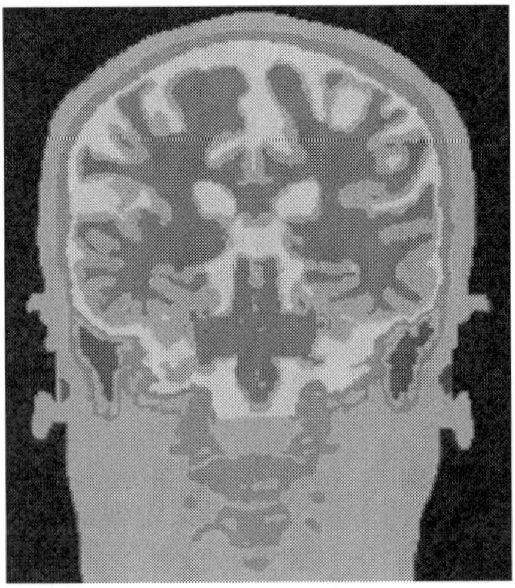

(b)

Figure 6: Skull Atlas-to-MRI registration. (a) Transformed MRI. (b) Fusion of Atlas skull and MRI segmentation.

The rigid transformation and the affine transformation use the same set of landmarks to formulate the similarity measures. Let Bm be the MRI brain mask defined as in MRI/CT registration. Let Ba be the Atlas brain mask defined in the same way as B_m. The landmarks are then the set of border voxels of B_m and the similarity measure is the sum of squares of distances from all landmarks to the contour of B_a. The coefficients of the rigid transformation are found by multiscale exhaustive searching while those of the affine transformation are determined by gradient-descent searching.

The source landmarks used in the thin-plate spline transformation include several parts. The two primary parts are the inferior source landmarks and the superior source landmarks. Inferior source landmarks are distributed over the inferior part of the contour of the Atlas brain mask B_a. The superior source landmarks are placed over the superior part of the contour of the Atlas head mask H_a. Their extraction is very similar to the extraction of those used in the MRI/CT registration, but they are much more sparsely distributed to reduce the computational overhead. The corresponding target landmarks in the subject head are determined automatically after the rigid and affine transformation are performed. For each superior source landmark, the corresponding superior target landmark is obtained as the intersection of the contour of the subject head mask Hm and the line between the superior source landmark and the subject brain center. We have two ways to obtain the inferior target landmarks according to their locations. The first method is the same as that obtaining the superior target landmarks and works for the posterior part of the inferior landmarks that are distributed around the cerebellum. For the second method that works for the anterior part of the inferior landmarks, we compute a field of distance from the contour of the brain mask B_m, and each inferior target landmark, which should be on the contour of Bm, is obtained by tracing from the corresponding source landmark following the gradient of the distance field.

In the procedure of the registration, the Atlas skull's thickness is adjusted according to an estimated ratio of the subject skull's thickness to the Atlas skull's thickness. This is achieved by using another set of source landmarks and another set of target landmarks. For each superior source landmark, we determine its "mate landmark" over the inner side of the Atlas skull and on the line between the superior source landmark and the Atlas brain center. For each superior target landmark, there is also a "mate landmark" on the line between the superior target landmark and the subject brain center. For each pair of the superior source landmark

S_i and its mate landmark S'_i, we have a distance dS_i; for each pair of the superior target landmark T_i and its mate landmark T' , we have a distance dT_i. A global ratio is then defined as dT_i / dS_i and its optimum is determined as the highest value so that the majority (such as 90%) of the superior target mate landmarks are off the mask Bm. For the few superior target mate landmarks that are within B_m according to the global ratio, the global value is decreased locally so that they are off the mask B_m. For each posterior inferior source landmark, we determine its mate landmark over the outside of the Atlas skull and on the line between the source landmark and the Atlas brain center. For each posterior inferior target landmark, we also have a mate landmark on the line between the target landmark and the subject brain center while its location is determined according to the global thickness ratio described above. In order to further improve alignment performance, there are few other landmarks such as the centers of the eyeballs. We have a specific algorithm to automatically detect the subject's eyeballs, but the details are not described here for the limit of space. When all source landmarks and all target landmarks are determined, the transformation for any point in the space can be calculated as shown in (2).

Registrations to the EEG Sensor Positions

The EEG sensor positions are automatically localized in 3D with the Geodesic Photogrammetry System (GPS). GPS captures images from 11 cameras in fixed positions to allow identification of the sensors in the images and then computation of the 3D coordinates (Figure 7). The Atlas-to-MRI-to-GPS registration then consists of two stages, the MRI-to-GPS registration and then the Atlas-to-MRI registration, using the newly transformed MRI. Atlas-to-MRI registration has been described in the above subsection, and this subsection only deals with the MRI-to-GPS registration. There are three steps. The first can be an automatic rigid transformation with other alternatives described below. The second is an affine transformation. The first two can bring the MRI to a good alignment state with the GPS data, but the GPS points may still be off the scalp. The third step then performs thin-plate spline transformation to deform the MRI so that each GPS point locates on the scalp. The landmarks used in the first two steps are the GPS points themselves, and the similarity measure is the sum of squares of the distances from the landmarks to the MRI scalp, that is, the contour of the head mask H_m. For the thin-plate spline transformation, the target landmarks are the GPS points while the source landmarks are those points on the contour of H_m so that each one is on the line connecting the brain center and the corresponding target landmark.

(a)

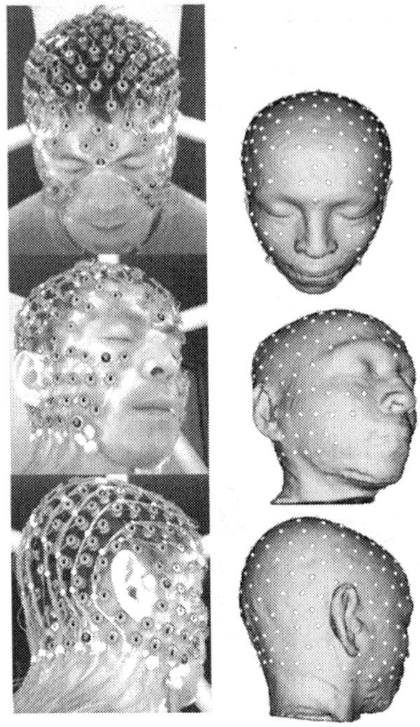

Figure 7: (a) Geodesic Photogrammetry System (GPS). (b) Identification of EEG sensors in the GPS software. (c) Registration of EEG sensors with the MRI head surface in BrainK.

The first automatic step of the MRI-to-GPS registration can be replaced by a transformation with the user providing three fiducial landmarks on the MRI space: the nasion and the left and right periauricular points (where the jaw hinge meets the skull). BrainK is able to stretch the MRI so that the user-specified landmarks match the corresponding ones in the GPS space. This initial state can be directly fed to the thin-plate spline transformation or it may pass through an intermediate affine transformation. BrainK also allows the user to specify only the nasion landmark and does rigid transformation in the first step while keeping the corresponding nasion landmarks matched.

The Atlas-to-GPS registration is very similar to the MRIto-GPS registration, except that the Atlas typically has greater geometric differences from the GPS than the subject's own MRI.The steps are similar to those for the MRI-to-GPS transformation just described.

For the MRI-to-GPS registration, however, if they are to be used, the fiducial landmarks have to be manually marked on the MRI, whereas the fiducials are included in the digital Atlas dataset for the Atlas-toGPS registration. In addition, one more fiducial landmark, the vertex, is used to accommodate the greater geometric difference between the Atlas and the subject.

The GPS-to-head registration can be seen as the inverse of the registration of MRI-to-GPS, but no thin-plate spline transformation is needed. The initial transformation, either automatic rigid transformation or the stretching in terms of fiducial landmarks matching, can put the GPS in a good initial state. If needed, an affine transformation can be performed to further improve the alignment. The final adjustment then moves each GPS point onto the scalp, if it is not on it yet, with an end result that is similar to finding the corresponding landmarks in the MRI-to-GPS registration. When GPS is not available, a set of generic sensor positions may be used, such as those from the appropriate Geodesic Sensor Net (channel count and size) selected from the database. This registration has two steps. First, the generic spherical sensor cloud undergoes a translation transformation so that the center of the sensor point cloud overlaps with the center of the brain. Second, in a procedure that matches the physical conformation of the geodesic tension structure to the head, each sensor is mapped to a point on the scalp by the ray emanating from the brain center to the sensor's position.

Cortical Surface Reconstruction and Inflation

The wellknown marching-cube isosurface algorithm [24] is performed on the cerebral cortex volume of each hemisphere to reconstruct the cortical surface mesh. In addition, BrainK supports cortical surface inflation by iteratively updating all points in the mesh, such that the new position of each point is the weighted sum of its old values with the weighted sum of the centroids of the triangles that are incident to the point.

Dipole Tessellation

The dipole tessellation operates in two modes. If the individual's MRI is available, the cortical surface is extracted and tessellated into patches, and a dipole is fit to each patch. If it is not available, then dipole triples (fitting x, y, z components of the unknown dipole orientation) are distributed evenly through the cortical gray matter. Whereas a trivial algorithm can distribute the triples, the even distribution of the oriented dipoles over the cortical surface is formulated as a sophisticated graph partition [25] problem. Given the cortical mesh from marching cubes, composed of a great number of interconnected tiny triangles, a graph is first constructed so that each vertex corresponds to a triangle and each edge incident to two vertices corresponds to the adjacency between the two corresponding triangles. The graph partition algorithm then divides the entire graph into a set of parts, with the constraints that (1) the mesh is divided into the same number of patches as the desired number of oriented dipoles, (2) all patches are very compact (the more like to a square or circle, the better), and (3) all patches have similar areas. The graph partition library Chaco [26] is used for the implementation. Given the mesh partition performed as above, an equivalent oriented dipole is then defined for each patch, generated at the center of the patch, with its orientation defined by the mean of all surface normals of the triangles in the patch, weighted by their areas. This equivalent dipole thus reflects the orientation of the electrical source if the entire patch was synchronously active, thereby serving as a useful approximation for the resolution of cortical activity that is set by the graph partitioning decision. The cortical surface and the oriented dipoles are displayed in Figure 8.

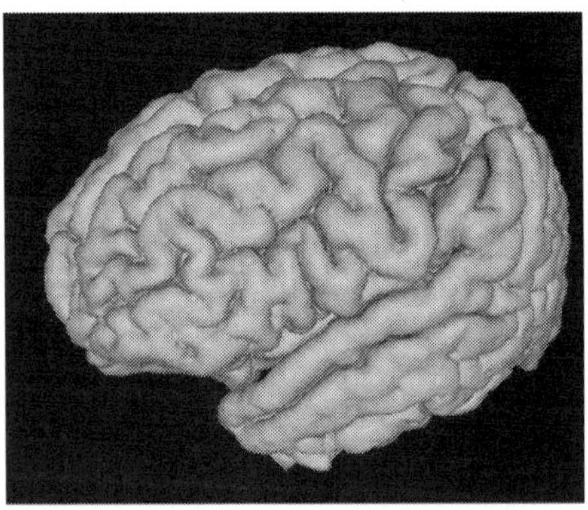

(a)

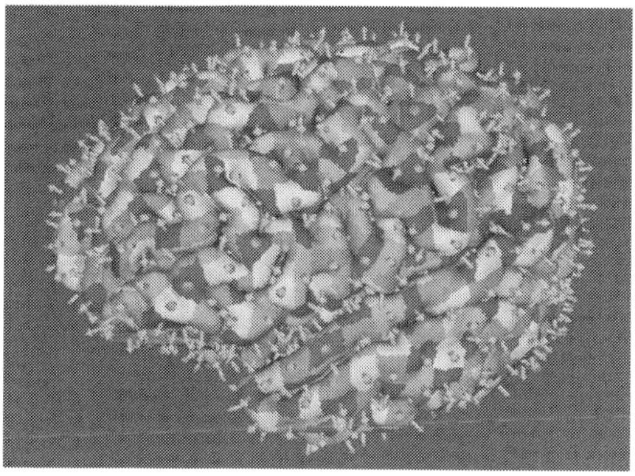

(b)

Figure 8: Cortical surface (a) and tessellation of the surface into ~1 cm^2 patches (about 1200 per hemisphere), each with an oriented equivalent dipole (b). Color is used here only to separate the cortical patches, such that every patch has a different color from its neighbors.

Talairach Transformation

The Talairach transformation component brings all the data, the segmentation, the cortical surfaces, the sensors, and the dipoles, into the standard Talairach space [27, 28]. The transformation is determined with three points specified by the user: the anterior commissure (AC), the posterior commissure (PC), and one point forming the middle plane with AC and PC. The procedure maintains the topology correctness of the cortical surface that is guaranteed in either the MRI segmentation component or the Atlas dataset and then maintained through each registration operation.

EVALUATION OF BRAINK'S TISSUE SEGMENTATION AND CORTICAL SURFACE EXTRACTION

BrainK provides certain algorithms, such as skull bone density (X-ray CT) image registration and cortical surface dipole tessellation, which are specifically important to model the electrical properties of the human head. As described below in Section 8, efficient construction of accurate electrical volume conduction head models is important to improve the source analysis of human electrophysiological activity with dense array technologies. Accurate head tissue conductivities and geometries are also important to optimize the transcranial current delivery to stimulate the brain that is now possible with dense array technologies.

In addition to these somewhat unique capabilities, BrainK implements certain tissue segmentation and surface extraction algorithms that are useful in medical image processing generally and that can be evaluated in comparison with existing software that has become well known in the neuroimaging research community. Examples of the software examined here are FreeSurfer, SPM, FSL, and BrainVisa. To assure accuracy for medical applications, validation of BrainK's unique algorithms against well-known reference software not only is scientifically useful but also forms an integral part of medical quality verification and validation. In addition, medical use requires that the functionality is fast and intuitive, such that BrainK must meet standards of efficiency and ease of use.

Our current evaluation work pays particular attention to brain tissue segmentation due to the fact that the accuracy of brain tissue segmentation is critical for a variety of medical and neurological applications such as source localization and cortical dysplasia detection.

Comparative Packages

Below is a brief description of the four brain image analysis tools used for comparative performance evaluation. Although these tools provide varying functionality in neuroimage analysis, they all support automatic T1-weighted human brain MR image segmentation. Our comparative evaluation is focused on the image segmentation task.

FreeSurfer

FreeSurfer [29, 30] is a set of tools for reconstruction of cortical surfaces from structural MRI data and for the overlay of functional data onto the reconstructed surface. FreeSurfer is developed in the Nuclear Magnetic Resonance (NMR) Center, Massachusetts General Hospital. The cortical surface reconstruction pipeline in FreeSurfer mainly consists of three steps. First, a brain mask is extracted with alignment of the structure MR image to the Talairach Atlas and the bias field is corrected. Then, the brain volume is labeled as various cortical or subcortical structures in a procedure based on both a subject-independent probabilistic Atlas and subject-specific measured values. Finally, the cortical surfaces are constructed from the prior segmentation, which involves a topology correction procedure.

SPM

SPM (Statistical Parametric Mapping) is a statistical technique for testing hypotheses about functional imaging data [31]. SPM also refers to the software developed by the Wellcome Department of Imaging Neuroscience, University College London, to carry out such analysis. SPM features structural MRI segmentation as well as a series of functional neuroimage analysis. Structural MRI segmentation in SPM can be characterized as a circular procedure that involves alternating three processing steps [32]: a bias correction step that corrects the intensity inhomogeneity, a registration step that normalizes the image to standard tissue probability maps, and a segmentation step that classifies image voxels into different tissue types. As the segmentation results, SPM assigns each image voxel three probabilities with respect to three tissue types: CSF, GM, and WM. Our experiments were conducted with SPM5, which is old with respect to the latest SPM12, but their implementation is based on the same algorithm presented in [32], although the newer version makes use of additional tissue classes and multichannel segmentation and incorporates a more flexible image registration component, as stated in the release notes of SPM12.

FSL

FSL (the FMRIB Software Library) is a collection of functional and structural neuroimage analysis tools [33]. For structural segmentation, FSL applies the Brain Extraction Tool (BET) for segmenting brain from nonbrain regions in structural and functional data, and FAST (FMRIB's Automated Segmentation Tool) for bias field correction and brain segmentation into three tissue types: CSF, GM, and WM. Structural MRI segmentation in FSL consists of two steps: using BET to extract the brain and using FAST to classify tissue types. BET performs skull stripping with a surface model [34]. The underlying method of FAST is based on an expectation-maximization algorithm combined with a hidden Markov random field (MRF) model [35]. Due to the regularization of the MRF model, FAST is supposed to be more robust to noise than standard finite mixture model based methods.

BrainVisa

BrainVisa [36, 37] is software developed at Service Hospitalier Frédéric Joliot (SHFJ) that encompasses an image processing factory and is distributed with a toolbox of building blocks dedicated to the segmentation of T1-weighted MR image. Structural MRI segmentation in BrainVisa consists of four main steps. First, the user prepares the data for segmentation by specifying several key landmark points including the anterior commissure (AC), the posterior commissure (PC), an interhemispheric point, and a left hemisphere point. A brain mask is then extracted including only white matter and gray matter integrating bias field correction [38] and histogram analysis [39]. This is followed by a hemisphere partition and removal of cerebellum with morphological image analysis [40]. Finally, cerebral gray matter and white matter are differentiated with histogram analysis [41].

DATASETS

The evaluation was performed on three image datasets: a set of BrainWeb data with ground truth segmentation, a set of IBSR data with manually guided expert segmentation, and a set of real scans of subjects with either mild cognitive impairment or Alzheimer's disease.

BrainWeb

The BrainWeb dataset is a group of 8 realistic T1-weighted MR simulated images with ground truth segmentation provided by BrainWeb, a simulated brain database [42, 43]. All 8 MR images are simulated on a normal anatomical model. The resolution of the images is 1 mm^3. In the ground truth image, all voxels in the image are segmented into the following tissue types: Background, CSF, GM, WM, Fat, Muscle/Skin, Skin, Skull, Glial Matter, and Connective.

A variety of noise levels and levels of intensity inhomogeneity (i.e., intensity nonuniformity (INU)) are artificially introduced in the simulated images, as listed in Table 1. As stated in BrainWeb documentation [44], the "noise" in the simulated images has Rayleigh statistics in the background and Rician statistics in the signal regions. The "percentage noise" number represents the percent ratio of the standard deviation of the white Gaussian noise versus the signal for a reference tissue. The noise reference tissue used in our dataset is white matter. The meaning of the intensity inhomogeneity level is as follows. "For a 20% level, the multiplicative INU field has a range of values of 0.90–1.10 over the brain area. For other INU levels, the field is linearly scaled accordingly (e.g., to a range of 0.80–1.20 for a 40% level)." According to BrainWeb, the INU fields are realistic in that they are slowly varying fields of a complex shape and were estimated from real MRI scans.

Table 1: Noise levels and IIH levels of the BrainWeb datasets.

Dataset	1	2	3	4	5	6	7	8
Noise level	3%	3%	5%	5%	7%	7%	9%	9%
IIH level	20%	40%	20%	40%	20%	40%	20%	40%

IBSR

The IBSR dataset is a group of 18 T1-weighted real MR brain datasets. Their manually guided expert segmentation is included in the Internet Brain Segmentation Repository (IBSR) supported by the Center for Morphometric Analysis (CMA) at Massachusetts General Hospital [45]. The slice resolution of all datasets is 1.5 mm and the XY resolution varies from 1 mm^2 to 0.837 mm^2. The MR images have been "positionally normalized" into the Talairach orientation, but all five tools performed on this group of data assumed that the images were not normalized. The MR images were also processed by the CMA bias field correction routines, but it is not guaranteed that the intensity inhomogeneity is completely corrected. All five tools therefore treated the datasets as if no bias field correction had been performed.

Each MR image was manually segmented into 44 individual structures including 3rd ventricle, 4th ventricle, Brainstem, and left and right: accumbens area, amygdala, amygdala anterior, caudate, cerebellum cortex, cerebellum exterior, cerebellum white matter, cerebral cortex, cerebral exterior, cerebral white matter, hippocampus, Inf Lat vent, lateral ventricle, pallidum, putamen, thalamus proper, ventral DC, and vessel.

The 18 MR images represent various levels of image quality. To organize the evaluation by general image quality, we divided the images into two subgroups: the first 13 MR images with good quality and 5 more MR images with bad quality. Note that the ordering of the IBSR datasets is different from the original order. A map of the order we used to the original order is "1, 2, 5, 6, 7, 8, 9, 10, 11, 12, 14, 15, 16, 17, 18, 3, 4, 13." For example, when we refer to the 3rd dataset, it is actually the 5th in the original order.

Datasets Reflecting Neuropathology

In addition to the BrainWeb and the IBSR datasets, which were used for both quantitative and qualitative evaluation, we also tested five tools on an auxiliary group of 8 real MR images scanned from subjects with minor cognitive impairment or those with Alzheimer's disease. The resolution of these datasets is $1.139 \times 1.211 \times 1.211$ mm^3. The source of these datasets is the Neurobiology Research Unit [46] in the University Hospital Rigshospitalet in Denmark. No ground truth or manual segmentation is provided for these datasets; the issue for the evaluation is primarily whether the abnormal brains can be segmented with apparently reasonable accuracy.

QUANTITATIVE EVALUATION

In this section, we present a quantitative evaluation on the segmentation accuracy, robustness, and computational efficiency of BrainK in comparison to other four packages. We use the widely used Dice metric [1, 4, 46, 47] as the measurement for segmentation accuracy. The standard deviation of the Dice metric provides a measure of segmentation robustness. Computational efficiency is measured as the required run time of each package.

Dice Metric

Let TP refer to the number of true positives, FP to false positives, and FN to false negatives. The Dice metric is then given by

$$\text{Dice metric} = \frac{2 \times \text{TP}}{2 \times \text{TP} + \text{FP} + \text{FN}} \tag{3}$$

TP, FP, and FN are measured versus the manual segmentation of real MR datasets or the ground truth of simulated images. Note that when the segmentation is given as a probability between 0 and 1 for each image voxel for each tissue class (such as in the case of SPM), then TP, FP, and FN are calculated as the sum of the probabilities instead of discrete counting.

For quantitative evaluation using Dice metric, we must decide the tissue type on which the metric is measured. BrainVisa only performs cerebrum segmentation while FSL and SPM segment the entire brain into CSF, GM, and WM without extraction of the cerebrum. FreeSurfer also performs segmentation on the whole brain, but it segments the brain into a greater number of tissue types (including cerebral white matter and cerebral cortex). Currently, BrainK performs segmentation on the entire image, and it is able to differentiate cerebrum from cerebellum and one cerebral hemisphere from the other. In our quantitative evaluation, we must calibrate the segmentation of the five packages within a standard framework, so that common tissue types can be used for quantitative metric measurements.

For the BrainWeb datasets, we calibrated the segmentation of five tools to the segmentation of cerebral WM and cerebral GM and measured the Dice metrics with respect to these two tissue types. To enable this, we manually partitioned the ground truth whole brain (WM plus GM) at the brainstem to extract the cerebral WM and the cerebral GM. Cerebral WM and cerebral GM also need to be extracted for the SPM and FSL segmentation results. We use a procedure (described in the next paragraph) that almost "perfectly" partitions the segmentation results based on the ground truth partition. For BrainVisa and BrainK, no transformation in the calibration is required. For FreeSurfer, we just need to relabel all cerebral cortex voxels and all subcortical voxels excluding cerebral WM as cerebral gray matter.

Let TP-Cerebrum and TP-Cerebellum, respectively, denote the set of true positives of cerebrum and cerebellum in the segmentation of SPM and FSL. Let FP-Brain denote the entire false positives including those in cerebrum and cerebellum. The partition of the brain segmented by FSL and SPM is essentially the partition of FP-Brain into false positives in cerebrum and those in cerebellum, which is described as follows. For each voxel V in FP-Brain, if it has a shorter path in FP-Brain to TP-Cerebrum than any paths in FP-Brain from V to TP-Cerebellum, then V is taken as a false positive (of WM or GM) in cerebrum; otherwise, it is taken as a false positive (of GM or WM) in cerebellum. For the IBSR data, we calibrated the segmentation of five tools to the segmentation of cerebral cortex and cerebral WM and measured the Dice metrics with respect to these two tissue types. These quantitative metrics give an evaluation on the accuracy of the cortical surface reconstruction which depend on segmentation of cerebral cortex and cerebral WM and are irrelevant to segmentation of subcortical gray matter tissues. Since FreeSurfer explicitly labels cerebral cortex and cerebral white matter, we do not need to do any transformation in the calibration. The calibration of FSL and SPM first conducts the brain partition to extract the cerebral WM and cerebral GM. Given the set of cerebral WM and cerebral GM segmented by FSL, SPM, BrainVisa, or BrainK, we measured the Dice metrics with respect to cerebral cortex and cerebral WM in the way described below.

Experiment Settings

FreeSurfer

We tested FreeSurfer on both the BrainWeb and the IBSR datasets in a fully automatic mode without any user intervention. An issue in collecting FreeSurfer segmentation results is the production of the cerebral cortex mask. There is a so-called "aseg" image and a "ribbon" image and both of these record voxels are labeled as cerebral cortex. The "ribbon" data is what FreeSurfer suggests to use, but it has more false negatives than the "aseg" data. In contrast, the "aseg" image is an intermediate result, and it has more false positives than the "ribbon" data. We applied a simple morphological closing operation on the union of the cortex ribbon and the subcortical structures so that certain true cerebral cortex voxels labeled in "aseg" but missed in "ribbon" are covered. This procedure apparently improved the performance of the cerebral cortex segmentation. The Dice metric for the "aseg" image was 0.7353, for the "ribbon" image was

0.8009, and for the "closed" image was 0.8212.

FSL

In our first batch of experiments with FSL, we let FSL automatically extract the brain and performed brain tissue classification on both the BrainWeb and the IBSR datasets. However, FSL generated poor results on the brain extraction and brain tissue classification on 6 of the IBSR datasets (dataset 5 to dataset 10). In our second batch of experiments, therefore, we used different parameters in FSL, obtained better brain masks for these datasets, and then repeated the brain tissue classification. The brain masks generated in the second batch of experiments were still not good enough. We therefore used the brain masks generated by FreeSurfer for the brain tissue classification in FSL. This delivered the best FSL brain segmentation performance on the 6 difficult IBSR datasets. The three batches of experiments on FSL show that the brain extraction algorithm of FSL is not robust on the IBSR datasets, but the brain tissue segmentation itself performed well given good brain masks. The performance of FSL on the 6 IBSR datasets with respect to the three batches of experiments is shown in Table 2.

Table 2: Dice metrics collected for FSL using different brain masks.

Brain masks	Tissue types	Dice metrics on 6 IBSR datasets					
		5	6	7	8	9	10
Default FSL Brain mask	Cerebral cortex	0.6591	0.6806	0.7268	0.6887	0.7713	0.6762
	Cerebral WM	0.8845	0.8928	0.8891	0.8335	0.9095	0.8792
Customized FSL Brain mask	Cerebral cortex	0.7608	0.7735	0.7772	0.7740	0.7854	0.7859
	Cerebral WM	0.8722	0.8747	0.8766	0.8711	0.8953	0.8767
FreeSurfer Brain mask	Cerebral cortex	0.7312	0.7559	0.7898	0.7587	0.8277	0.7471
	Cerebral WM	0.8862	0.8953	0.8912	0.8914	0.9146	0.9009

SPM

In our first batch of experimental tests with SPM, we used the default parameters and let SPM automatically perform brain tissue segmentation on the BrainWeb and the IBSR datasets. In the second batch of experiments, we changed the parameter "bias regularization" from the default "very light regularization" to "medium regularization" and reran SPM on the IBSR datasets. SPM is supposed to be used with greater bias regularization when it is known a priori that there is less intensity inhomogeneity in the image. Since the IBSR datasets were processed with bias field correction, the use of "medium regularization" rather than the default "very light regularization" improved the performance of SPM on almost all IBSR datasets. For the cerebral cortex, the mean Dice metric over the 18 ISBR datasets was 0.7609 for very light regularization and 0.7656 for medium regularization. The mean Dice metric for white matter was 0.8572 for very light regularization and 0.8707 for medium regularization. We used the best regularization for each dataset in our comparative evaluation.

BrainVisa

We tested BrainVisa on both the BrainWeb and the IBSR datasets automatically, with the exception that we manually specified landmark points including the AC point, the PC point, an interhemispheric point, and a left hemisphere point. BrainVisa produced an empty brain mask in the 9th IBSR dataset and was unable to generate brain masks for the 13th and the 18th datasets. In both cases, we set the Dice metrics to be 0.

BrainK

BrainK was tested on the BrainWeb and the IBSR datasets fully automatically.

Comparison Results

Segmentation Accuracy on the IBSR Datasets

We collected the Dice metrics with respect to cerebral cortex (Table 3) and cerebral WM (Table 4) using the five software packages on the IBSR datasets.

Table 3: Dice metrics of five tools with respect to cerebral cortex on the IBSR datasets.

IBSR datasets	Dice metrics with respect to cerebral cortex				
	BrainVisa	SPM5	FreeSurfer	FSL	BrainK
1	0.7461	0.7705	0.8039	0.7803	**0.8682**
2	0.7953	0.8048	0.8175	0.8121	**0.8619**
3	0.7674	0.8080	0.8362	0.8361	**0.8714**
4	0.7233	0.8356	**0.8641**	0.8028	0.8612
5	0.2875	0.4621	0.7794	0.7312	**0.8638**
6	0.6610	0.6609	0.8068	0.7559	**0.8441**
7	0.7108	0.7065	0.7888	0.7898	**0.8638**
8	0.6982	0.7670	0.7800	0.7587	**0.8790**
9	0	0.7595	0.8128	0.8277	**0.8700**
10	0.7707	0.6868	0.7729	0.7471	**0.8611**
11	**0.8688**	0.8396	0.8702	0.8833	0.8634
12	0.8596	0.8541	0.8458	0.8582	**0.8772**
13	0	0.8588	0.8647	0.8554	**0.8673**
14	0.8406	0.8426	0.8065	**0.8429**	0.8315
15	0.8441	0.8439	**0.8487**	0.8381	0.8457
16	0.8260	0.8263	0.8413	**0.8426**	0.8281
17	**0.8445**	0.8443	0.8076	0.8278	0.8379
18	0	**0.8539**	0.8337	0.8278	0.8379
Mean	0.6247	0.7656	0.8212	0.8121	**0.8557**
Mean on 13 good datasets	0.6068	0.7365	0.8187	0.8030	**0.8656**

Table 4: Dice metrics of five tools with respect to cerebral WM on the IBSR datasets.

IBSR datasets	Dice metrics with respect to cerebral WM				
	BrainVisa	SPM5	FreeSurfer	FSL	BrainK
1	0.8652	0.8927	0.7964	**0.8971**	0.8789
2	0.8899	0.8940	0.8208	**0.9160**	0.8926
3	0.8649	0.8936	0.8138	**0.9084**	0.8874
4	0.8596	0.9013	0.8489	**0.9168**	0.9070
5	0.4097	0.7312	**0.9240**	0.8862	0.9101
6	0.7970	0.7422	**0.9115**	0.8953	0.8954
7	0.8255	0.8734	**0.9147**	0.8912	0.9001
8	0.8110	0.8923	**0.9203**	0.8914	0.9170
9	0	0.9059	0.9179	0.9146	**0.9224**
10	0.8457	0.8744	**0.9069**	0.9009	0.8790
11	0.8975	0.8927	0.8711	**0.9142**	0.8969
12	0.8858	**0.9014**	0.8099	0.8988	0.9006
13	0	**0.8955**	0.8647	0.8673	0.8748
14	0.8613	**0.8678**	0.7824	0.8632	0.8622
15	0.8541	0.8880	**0.8746**	0.8743	0.8637
16	0.8792	**0.8948**	0.8479	0.8933	0.8713
17	**0.8619**	0.8551	0.8592	0.8585	0.8463
18	0	**0.8764**	0.8124	0.8323	0.8463
Mean	0.6893	0.8707	0.8610	**0.8900**	0.8862
Mean on 13 good datasets	0.6886	0.8685	0.8708	**0.8999**	0.8971

On average, BrainK performed best on cerebral cortex segmentation on all 18 IBSR datasets, good and bad, and on the 13 good datasets exclusively. In particular, BrainK's cerebral cortex performance is consistently better than the four other packages on the 13 good datasets except for the 4th dataset, where BrainK's performance is almost identical to the best, and the 12th dataset, where BrainK's performance is close to the best. The cerebral cortex segmentation performance of the five packages on the five bad datasets is similar, except that BrainVisa generated empty brain mask for the 18th dataset.

To provide an estimate of the statistical significance of the comparison of BrainK with the other packages, we conducted paired comparison t-tests across these 18 datasets. BrainK was more accurate at segmenting cerebral cortex with the ISBR datasets than BrainVisa ($p < 0.02$), SPM ($p < 0.006$), FreeSurfer ($p < 0.0005$), and FSL ($p < 0.0009$). If we omitted the three datasets on which BrainVisa failed, the comparisons again showed BrainK to be more accurate than BrainVisa ($p < 0.002$), SPM ($p < 0.009$), FreeSurfer ($p < 0.001$), and FSL ($p < 0.002$).

On average, FSL performed best on cerebral WM segmentation on the 18 IBSR datasets and on the 13 good datasets. However, BrainK's performance was very close to FSL's performance in this comparison. The performance of the five packages on the five bad datasets is similar, except that BrainVisa generated empty brain mask for the 18th dataset, and FreeSurfer gave particularly poor performance for the 14th dataset.

Segmentation Robustness on the IBSR Datasets

We calculated the standard deviations of the Dice metrics on the IBSR datasets. Together with the mean Dice metrics, these values indicate the segmentation robustness of the five packages on MR images scanned from different subjects. Greater mean Dice metric and lower standard deviation indicate greater robustness with respect to segmentation accuracy.

Two groups of standard deviations were calculated: one on the total 18 IBSR datasets and one on the 13 good IBSR datasets. BrainK demonstrated the lowest standard deviation of the software packages with respect to cerebral cortex on both the total 18 IBSR datasets and the 13 good IBSR datasets, as shown in Table 5. The lowest mean and standard deviation Dice metric with respect to cerebral cortex indicate that BrainK demonstrates the best accuracy and robustness with respect to cerebral cortex on the IBSR datasets. For cerebral WM, BrainK and FSL performed very similarly with respect to both the mean and the standard deviation of the Dice metric on both the total 18 IBSR datasets and the 13 good IBSR datasets. BrainK and FSL tied for the best accuracy and robustness with respect to cerebral WM on the IBSR datasets. Considering both cerebral cortex and cerebral WM, these results suggest that, in general, BrainK demonstrated the best overall performance on the IBSR datasets.

Table 5: Standard deviation of Dice metrics of five tools on the IBSR datasets.

Sample groups		Standard deviations				
Datasets	Tissue type	BrainVisa	SPM5	FreeSurfer	FSL	BrainK
All IBSR datasets	Cerebral cortex	0.3155	0.1284	0.0308	0.0429	0.0192
	Cerebral WM	0.3351	0.0505	0.0473	0.0230	0.0225
13 good datasets	Cerebral cortex	0.3047	0.1414	0.0345	0.0476	0.0087
	Cerebral WM	0.3305	0.0593	0.0482	0.0143	0.0147

Segmentation Robustness with respect to Noise and IIH on the BrainWeb Datasets

As described in Section 3.1, the BrainWeb datasets vary in noise levels and intensity inhomogeneity (IIH) levels. The performance in Dice metrics of the five packages with respect to cerebral GM and cerebral WM is listed in Tables 6 and 7.

Table 6: Dice metrics of five tools with respect to cerebral GM on the BrainWeb datasets.

BrainWeb datasets		Dice metrics with respect to cerebral GM				
Noise level	IIH level	BrainVisa	SPM5	FreeSurfer	FSL	BrainK
3%	20%	**0.9292**	0.9173	0.8333	0.9242	0.9084
	40%	0.9247	0.9189	0.8342	**0.9268**	0.9086
5%	20%	**0.9197**	0.8989	0.8323	0.9193	0.8908
	40%	**0.9201**	0.8998	0.8323	0.9193	0.8858
7%	20%	0.8628	0.8673	0.8320	**0.9113**	0.8816
	40%	0.8740	0.8713	0.8312	**0.9127**	0.8827
9%	20%	0.8166	0.8255	0.8259	**0.8996**	0.8658
	40%	0.7836	0.8301	0.8264	**0.9019**	0.8678

Table 7: Dice metrics of five tools with respect to cerebral WM on the Brain-Web datasets.

BrainWeb datasets		Dice metrics with respect to cerebral WM				
Noise level	IIH level	BrainVisa	SPM5	FreeSurfer	FSL	BrainK
3%	20%	0.9550	0.9471	0.8849	**0.9672**	0.9558
	40%	0.9599	0.9533	0.8889	**0.9664**	0.9593
5%	20%	0.9552	0.9314	0.8824	**0.9567**	0.9494
	40%	0.9534	0.9315	0.8863	**0.9581**	0.9476
7%	20%	0.9325	0.8978	0.8779	**0.9448**	0.9382
	40%	0.9311	0.9008	0.8796	**0.9467**	0.9370
9%	20%	0.8926	0.8656	0.8757	**0.9332**	0.9296
	40%	0.8748	0.8701	0.8740	**0.9354**	0.9289

Among the five packages, FreeSurfer demonstrated the lowest performance variation over different noise levels. SPM and BrainVisa are similar in showing the highest performance variations over different noise levels. BrainK and FSL have medium performance variations over different noise levels, compared to the other three packages. Although FreeSurfer performed consistently over different noise levels, it also gave results with the lowest accuracy on average.

For each of the four noise levels, we also tested the packages on images with two different IIH levels. All five packages gave little variation over different IIH levels. The only exception is for BrainVisa with N=9% and IIH = 40%. This is due to a poor brain mask.

It is worth noting that, in real MR scans, the intensity inhomogeneity may occur in various and unknown patterns. Such noise could occur together with other difficulties that may be not present in the simulated BrainWeb datasets. Therefore, we remark that our experiments with the BrainWeb dataset should not be taken to give a thorough and sufficient evaluation on the five packages with respect to the IIH robustness.

Computational Efficiency

The execution times of the fives packages tested on the IBSR datasets and the BrainWeb datasets are listed in Table 8. The experiments were all run on a single 2.8 Ghz Intel Xeon processor. Among the five packages, BrainVisa took the least amount of time, but it also produced the lowest segmentation accuracy and robustness on the IBSR datasets. FreeSurfer is well known for long execution times. However, these long execution times cover segmentation of more subcortical structures as well as reconstruction of cortical surfaces. It should be noted that most of the BrainK time was spent for topology correction, which was not included in the execution times of the other three.

Table 8: Computation times of five tools on the IBSR and the BrainWeb datasets.

Datasets	Computation times					
	BrainVisa	SPM5	FreeSurfer	FSL	BrainK	BrainK (topology correction)
IBSR	1.5 m	34 m	27.2 h	5 m	17 m	14 m
BrainWeb	1.6 m	20 m	24.5 h	9 m	21 m	18 m

QUALITATIVE EVALUATION

In this section, we give a qualitative evaluation of the five packages based on the results of applying the packages to the IBSR datasets, the BrainWeb datasets, and the 8 pathological datasets with mild cognitive impairment or Alzheimer's disease. We first summarize and compare the segmentation functionalities of the five packages followed by the discussion of their automaticity. Finally, we present various segmentation abnormalities that we observed in the experiments.

Segmentation Functionalities

The following is a summarization and comparison on the main segmentation features of the five packages:(i)Bias Field Correction. Freesurfer, SPM, FSL, and BrainVisa all integrate a bias field correction procedure, either prior to tissue classification or combined with the classification. BrainK, on the other hand, does not need an explicit bias field correction because the relative thresholding method is robust to bias field in arbitrary patterns.(ii)Brain Extraction. FSL, FreeSurfer, and BrainVisa provide separate tools for brain extraction (i.e., skull stripping) prior to brain tissue classification, whereas SPM combines brain extraction with tissue classification. BrainK, on the other hand, performs brain extraction after tissue classification. Note that the brain mask generated by BrainVisa is meant to contain only GM and WM while the brain mask generated by FSL and FreeSurfer is meant to contain CSF as well as GM and WM.(iii)Tissue Classification. FSL and SPM segment the brain volume into three tissue types: CSF, GM, and WM. BrainVisa extracts cerebral WM and cerebral GM. BrainK classifies the entire region of interest into CSF, GM, and WM, extracts the GM and WM, and separates cerebellum from cerebrum. BrainVisa and BrainK also provide cerebral hemisphere partition. FreeSurfer segments a whole brain into 37 individual structures including cerebral cortex, cerebral WM, a set of subcortical structures, brainstem, and cerebellar structures.(iv)Cortical Surface Reconstruction. BrainVisa, FreeSurfer, and BrainK support cortical surface reconstruction, but FSL and SPM do not. A core requirement for cortical surface reconstruction is to insure that the topology of the cortical surface is correct.(v)Multichannel Segmentation. It is worth noting that FSL and SPM12 provide multichannel segmentation such that T1- and T2-weighted images can be combined together for increased accuracy. However, T2-weighted images are often not available and when they are available, T2-weighted images often decrease the contrast between grey and white matter and hence lead to bad performance in practice.

Segmentation Automaticity and User Intervention

All five packages support highly automatic brain segmentation, with little or no user intervention. FreeSurfer allows the user to start the cortical surface reconstruction without any intervention. In case the segmentation is not satisfactory, FreeSurfer supports interactive tools to allow the user to modify the brain mask and to add control points to improve the intensity normalization of WM, a procedure that is extremely important for the performance of FreeSurfer's whole brain segmentation. FreeSurfer also supports interactive tools for editing the final results generated by the automatic processing.

FSL also allows the user to start the segmentation without any intervention. In FSL, brain extraction and tissue classification are performed, respectively, by the BET and FAST algorithms. The BET performance substantially influences that of FAST. If the user is not satisfied with the brain extraction, FSL allows the user to select different parameters and rerun BET. However, our experiments with BET on the IBSR datasets and the pathological datasets show that BET cannot guarantee good brain extraction even with user intervention. FAST has custom options for the user to select whether to use the k-means segmentation or a priori probability maps for initial segmentation and to guide the k-means segmentation with manual intervention.

In SPM, brain segmentation can also be automatically started with the default parameters and SPM often generates good results. An important custom parameter of SPM is the one that controls the extent of bias field regularization. When any parameter is changed, the segmentation procedure has to be started over from scratch.

BrainVisa requires the user to prepare the data by first specifying several landmark points including the AC point, the PC point, an interhemispheric point, and a left hemisphere point. When the data is prepared by the user, BrainVisa automatically performs segmentation. BrainVisa supports interactive tools for the user to edit the segmentation results.

BrainK performs segmentation fully automatically, but the user also has the opportunity to tune the performance by changing key parameters such as the relative thresholds. Compared to the user intervention mechanisms in the other four packages, user intervention in BrainK is in the form of global parameter selection and has the following advantages. First, it is straightforward and requires little or no expertise to understand the meaning of the parameters and the criterion for selecting optimal ones. Second, it is very easy to operate by sliding a value bar. Third, it is very efficient and the user can obtain the effect of parameter selection in real time for such parameters as relative thresholds. Fourth, the parameters have global effect for segmentation and the user does not need to repeat similar operations for different local regions.

Segmentation Abnormalities

In examining the quality of the results, we first attempted to understand why BrainK consistently achieved better performance than the other packages with respect to cerebral cortex segmentation on the 13 good IBSR datasets [48]. We found that there was a "shrinking" effect on the cerebral cortex segmentation for FreeSurfer, FSL, SPM, and BrainVisa, which gives rise to a significant number of false negatives. This problem did not occur or was much milder in BrainK. We think the underlying reason is that BrainK uses a new image modeling mechanism that can adapt to wider variations of GM intensities, whereas the statistical methods used in other packages were misled by these intensity variations, resulting in missing many GM voxels that had lower intensities.

The contrasting cortex segmentation of the five packages is shown for a representative IBSR dataset in Figure 9. Pink stands for correct WM, red for false WM negative, light green for correct GM, very light green for false WM negative and false GM negative, dark green for false GM positive, and gray false for GM negative. Note that, for SPM, light green represents correct GM segmentation, darker green represents false GM negative, and gray level represents false GM positive.

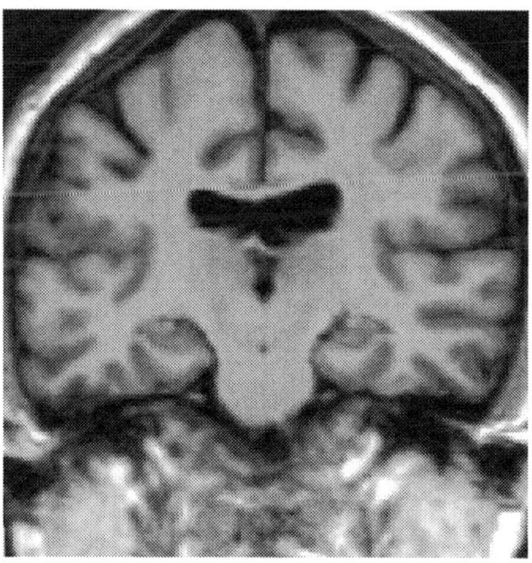

(a) MR image

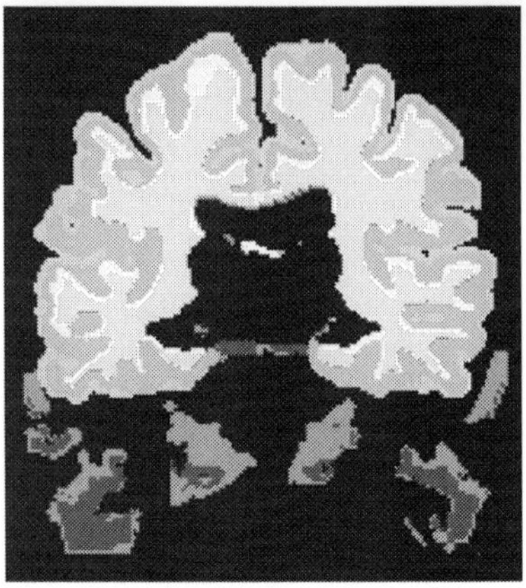

(b) Brain Visa result

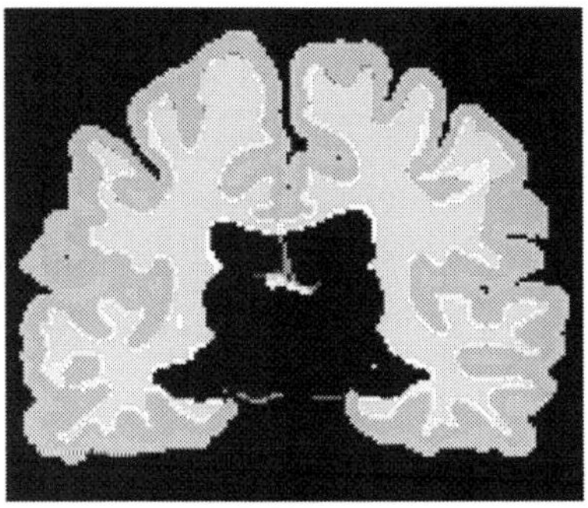

(c) Free Surfer result

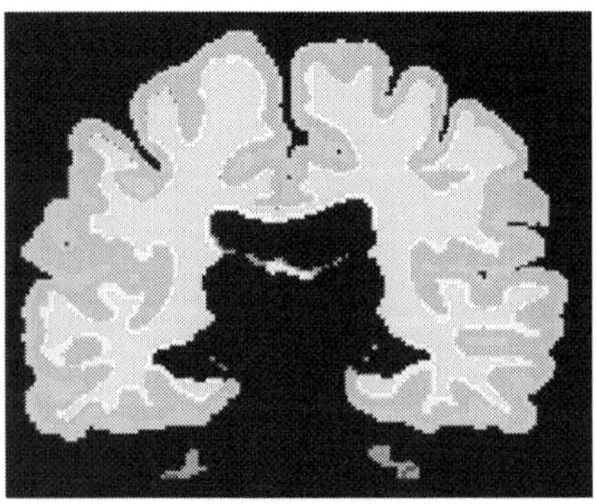

(d) FSL result

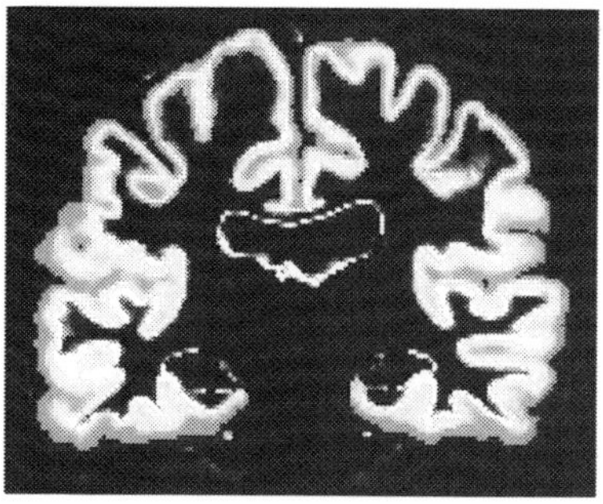

(e) SPM result

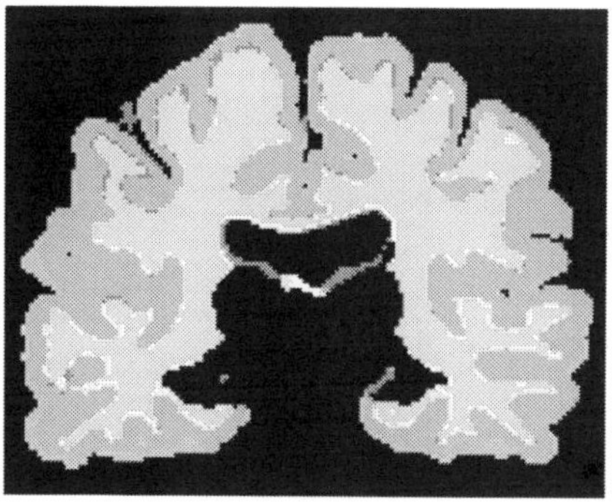

(f) BrainK result

Figure 9: GM-shrinking phenomenon.

A common problem in FSL, BrainVisa, and SPM is the poor brain extraction. Some examples are shown in Figure 10. Poor generation of the brain mask was the primary factor in the very poor performance for SPM and BrainVisa on the IBSR datasets. Since we used the considerably better brain masks generated by FreeSurfer for the FSL analysis, the FSL analysis is not representative. FreeSurfer did not encounter poor skull stripping, but an inaccurate brain mask may still be generated, as shown in Figure 11(b), where some nonbrain voxels with high intensities are taken as brain tissues. BrainK, on the other hand, does not depend on a brain extraction preprocessing step and robustly generated clean cerebrum masks as the union of the cerebral white matter and the cerebral gray matter on all tested datasets.

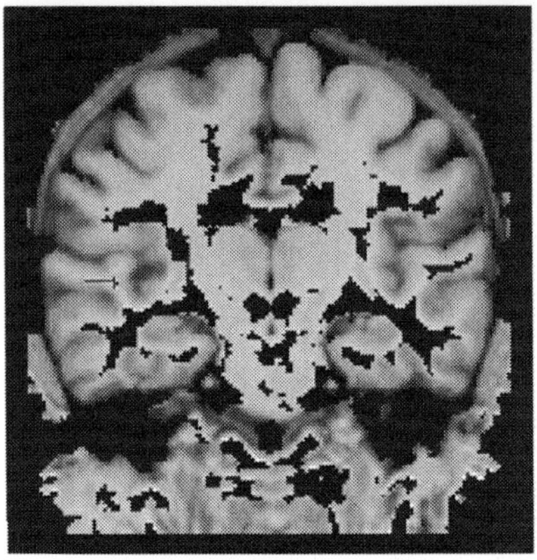

(a) Brain Visa brain mask

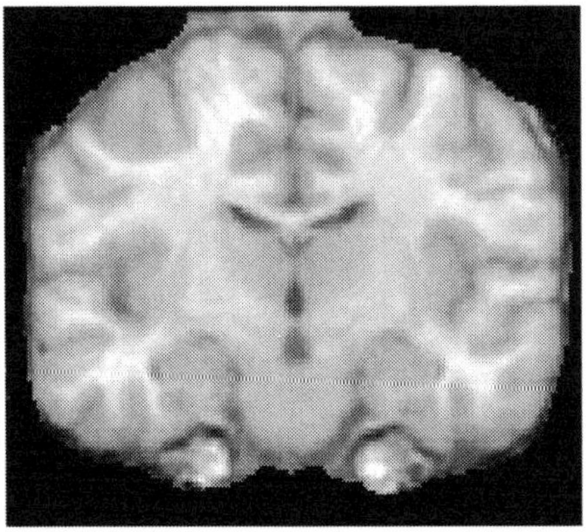

(b) FreeSurfer brain mask

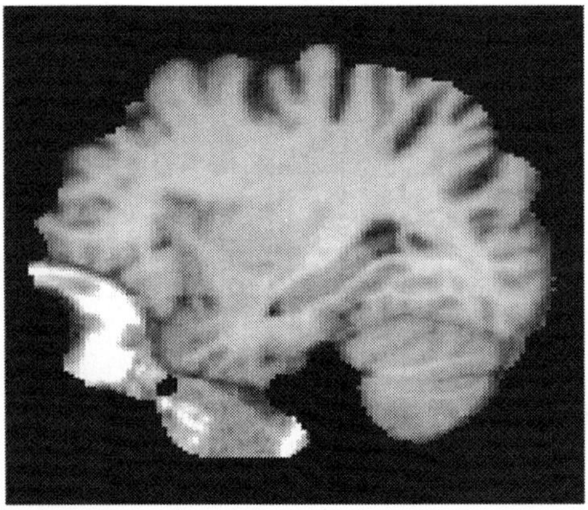

(c) FSL brain mask

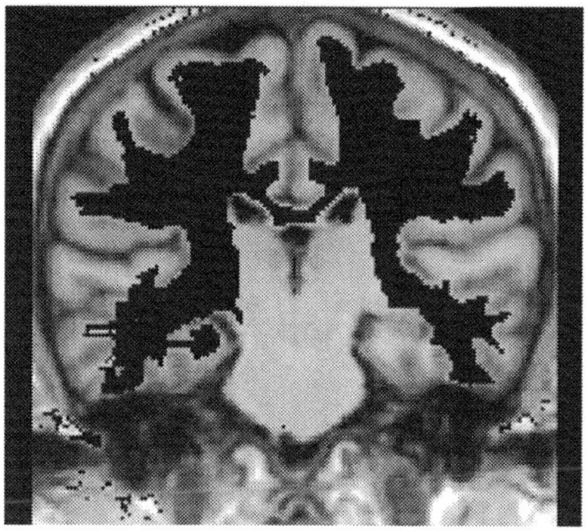

(d) SPM GM mask

Figure 10: Poor brain masks.

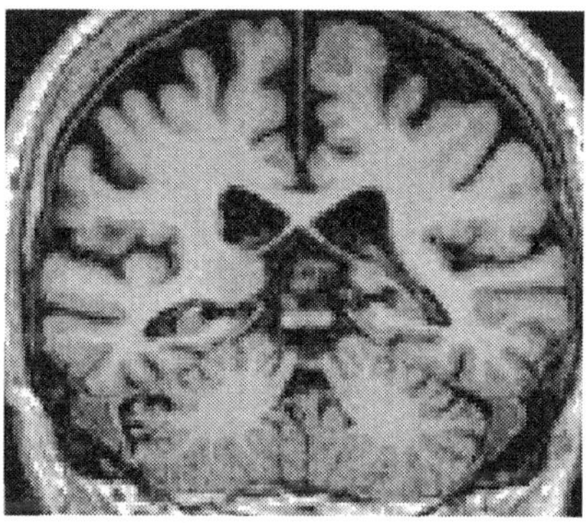

(a) MR image

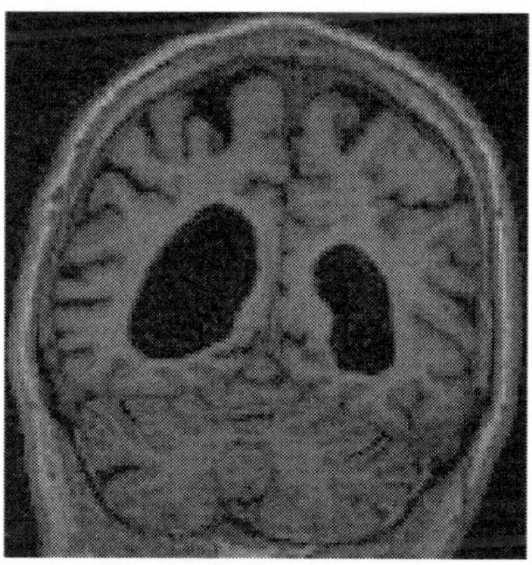

(b) MR image

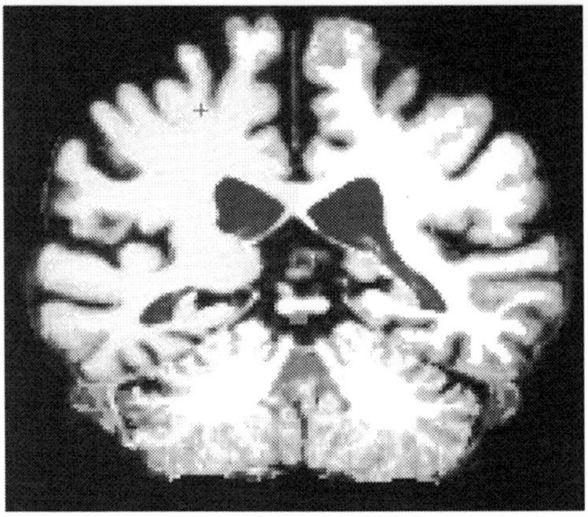

(c) FreeSurfer result for MRI in (a)

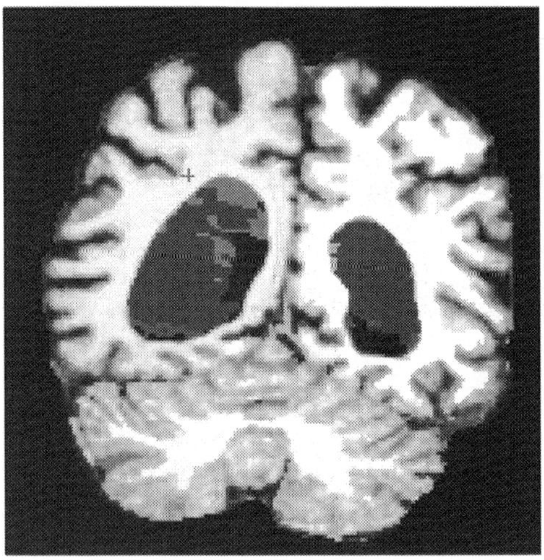

(d) FreeSurfer result for MRI in (b)

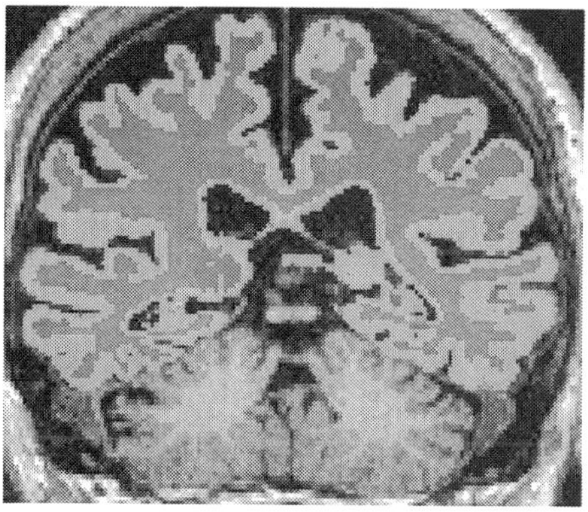

(e) BrainK result for MRI in (a)

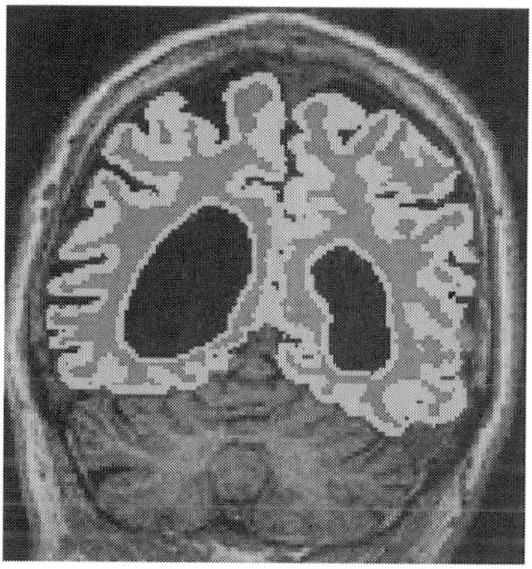

(f) BrainK result for MRI in (b)

Figure 11: FreeSurfer abnormalities on the pathological datasets compared to BrainK.

We also found some other interesting abnormalities with FreeSurfer, as shown in Figure 11. For example, in Figure 11(c), FreeSurfer cut off a significant amount of cerebral WM and cortex at the top of the brain. In Figure 11(d), FreeSurfer was unable to segment the complete lateral ventricle of the subject with Alzheimer's disease. In Figure 12, FreeSurfer generated poor GM/WM segmentation even for a simulated image with excellent quality (noise level is 3% and IIH level is 20%) while BrainK and FSL generated excellent results. These abnormalities, we believe, are due to the overregularization of the a priori probability maps used in FreeSurfer.

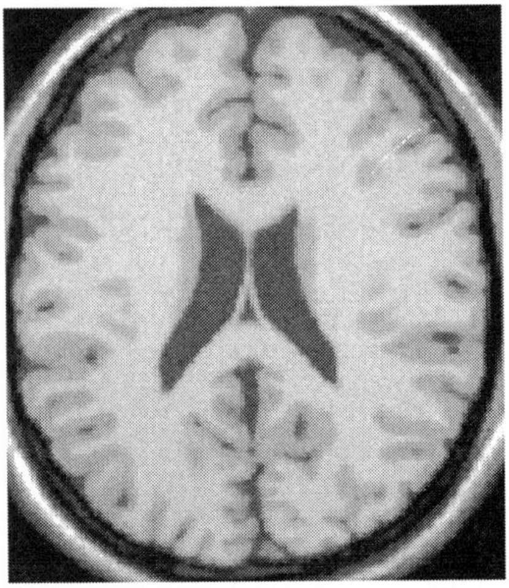

(a) BrainWeb MRI: N = 3% and IIH = 20%

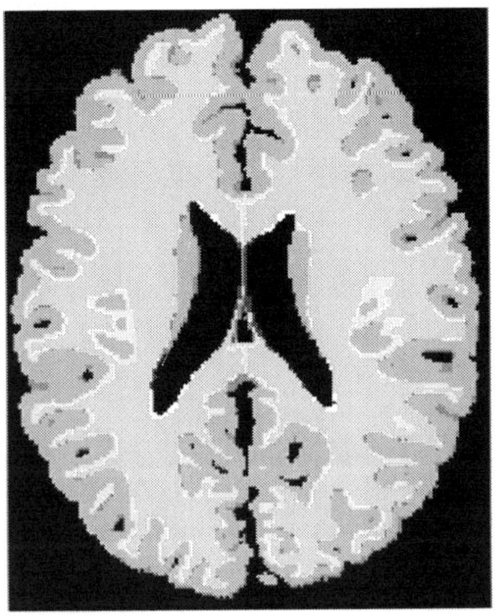

(b) FreeSurfer result

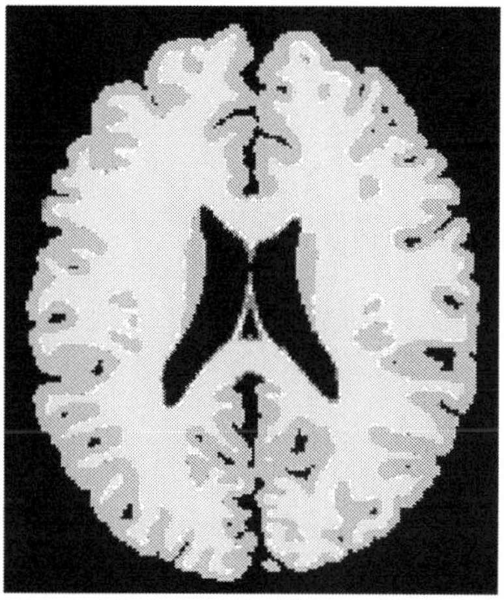

(c) FSL result

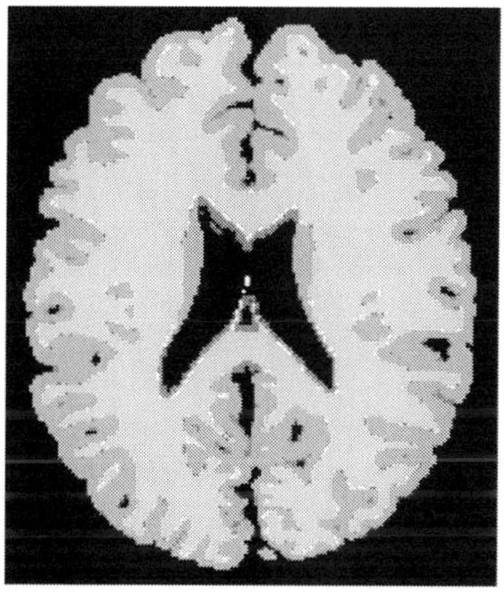

(d) BrainK result

Figure 12: BrainK segmentation on the BrainWeb MRI compared to FreeSurfer and FSL.

APPLICATIONS

At this point, BrainK has built functional electrical head models for more than 200 persons. Critical evaluation of BrainK's performance for generating segmented volumes and cortical surfaces continues in the applications of the Modal Image Pipeline, EGI's commercial software that incorporates the BrainK algorithms and workflows. Validation tests of physical tissue boundaries are conducted in the routine workflow for each head model construction. A first step in the FDM electrical lead field generation [49, 50], EGI's head physics code, is to test whether there are physically realistic segmentation properties in the BrainK output required for accurate electrical modeling. For example, there must be at least one voxel of CSF between the outer gray matter of the cortex and the inner table of the skull.

There have been many applications of BrainK in research and development projects that have tested its capabilities and limitations. For example, although the skull Atlases are organized by age and gender in a database to minimize the warping required to fit an individual's MRI, tests have been conducted with skull CTs and MRIs of divergent shape (Figure 13), with good results.

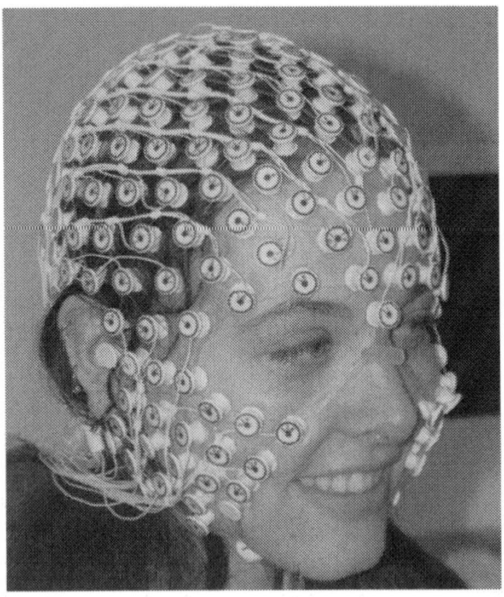

(a)

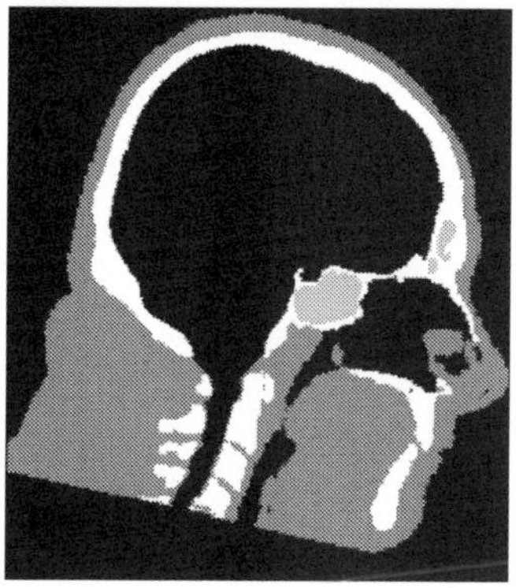

(b)

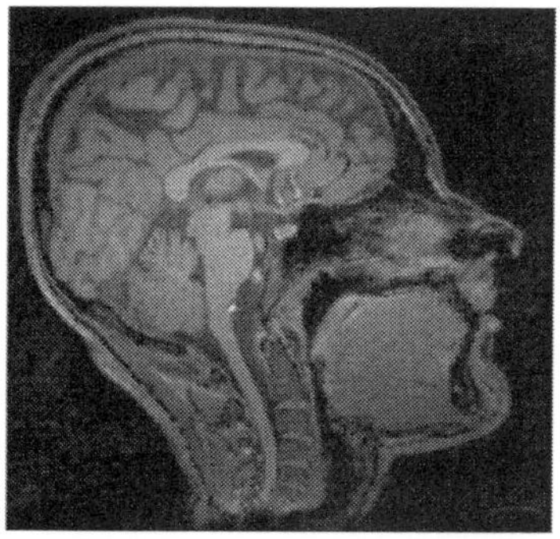

(c)

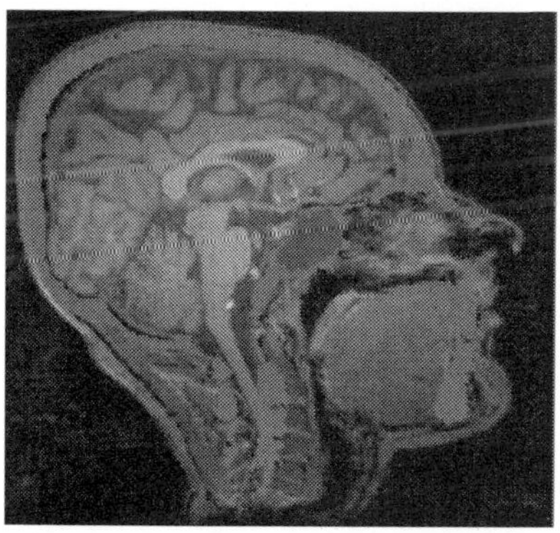

(d)

Figure 13: Skull Atlas-to-MRI registration with extensive warping of the skull to fit. (a) A volunteer in a 256 Net. (c) This person's MRI. (b) A database Atlas skull with a different shape. (d) Warp of the Atlas database skull to the volunteer's MRI.

Electrical source analysis with children requires age-appropriate electrical head models. Infants particularly require unique models because of their small head size and highly conductive skulls. BrainK has proven effective in adapting to the unique challenges of processing infant MRIs and CTs. Figure 14 shows results from a current project to create pediatric electrical head models directed by Sergei Turovets at EGI and by Linda Larsen-Prior and Fred Prior at University of Washington St. Louis. Figure 14(a) shows a normal infant's segmented MRI, with that infant's CT, registered to the segmented MRI, showing the thin skull and fontanel anterior to the vertex. Figure 14(b) shows the extracted cortical surface, and Figure 14(c) shows the dipole tessellated cortical surface, prepared for electrical source analysis.

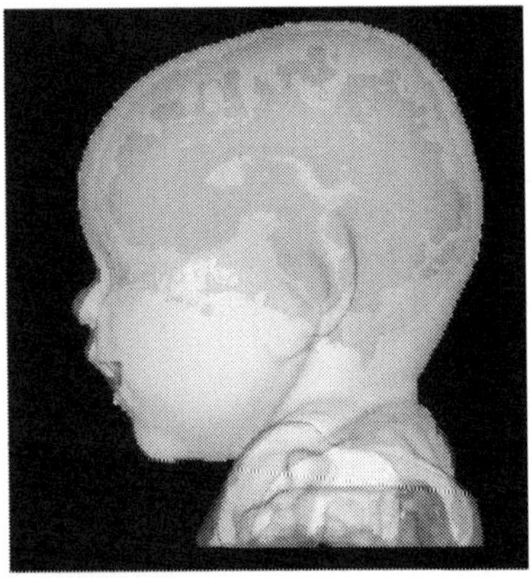

(a)

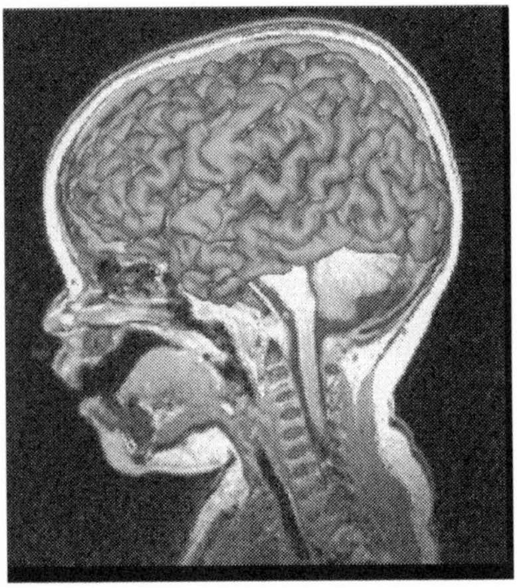

(b)

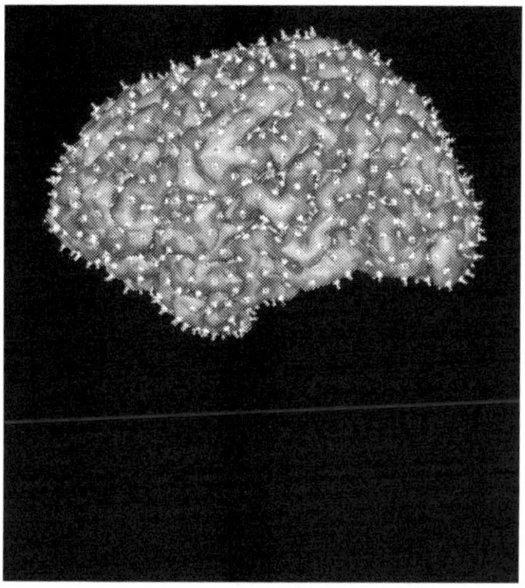

(c)

Figure 14: Electrical head model for an infant. (a) Registration of skull from CT with segmented MRI. (b) Cortical surface extraction overlaid with MRI. (c) Dipole tessellation of cortical surface for electrical source analysis.

The FDM electrical head models created by BrainK allow accurate estimation of the delivery of current, as well as its measurement, from the dense electrode array. With BrainK's anatomical results providing the geometry for the head physics code (Poisson equations for electromagnetic propagation in a finite difference model or FDM), lead fields are generated in the Modal Image Pipeline head physics code, describing the propagation from cortical sources to head surface electrodes (for EEG analysis) and from head surface electrodes to cortical sources (for transcranial electrical neuromodulation). The results are visualized on BrainK's anatomy and registered electrode positions in the reciprocity engine developed by Erik Anderson of EGI (Figure 15). In this illustration, cortical sources were chosen in the approximate region of auditory cortex, with the equivalent dipole for each cortical patch visualized with a gold arrow (seen best in Figure 15(b)). The lead field computation with the head physics model then shows the resulting head surface fields in the color of both the scalp surface and the electrodes (red is positive; blue is negative). Given this generated surface field, the reciprocity theorem states that impressing current on those cortical sources can be accomplished by applying current in the same surface pattern (red for anodal and blue for cathodal). The lead field from the head physics code is again used to estimate the current density that is impressed on the cortex by electrical neuromodulation with this electrode pattern (Figures 15(c) and 15(d)). The accuracy of the electrical solutions is fully dependent on the detailed geometry provided by BrainK, including not only the accurate (<1 mm) registration of sensors to the head surface, but also the specification of the orientation of the cortical patches provided by BrainK's characterization of the cortical surface.

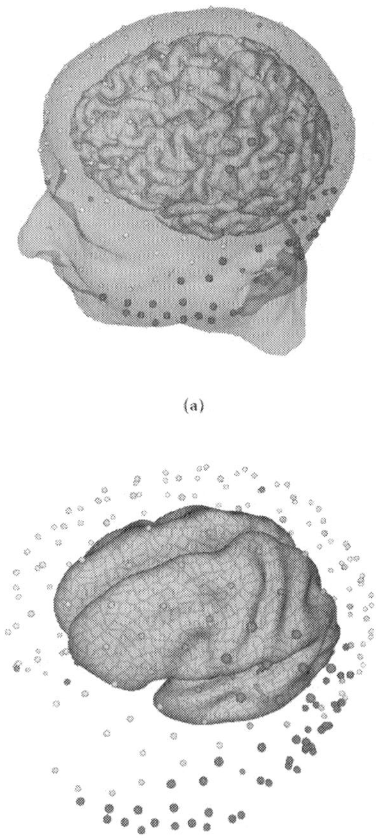

(a)

(b)

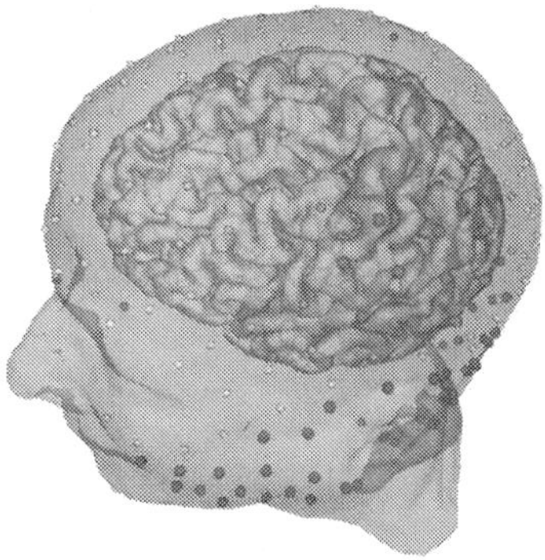

(c)

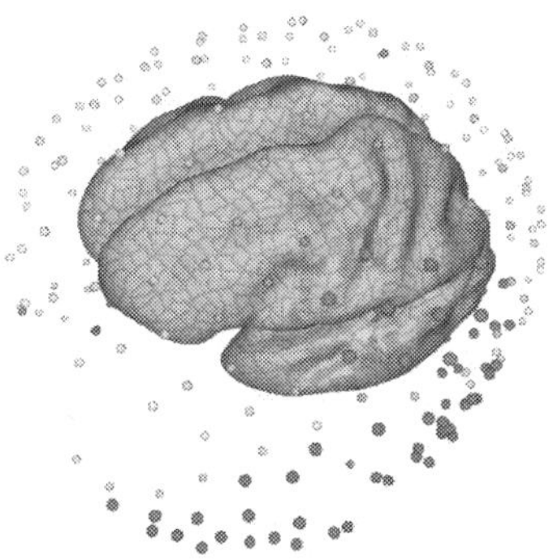

(d)

Figure 15: ((a), (b)) Using BrainK's anatomy in the reciprocity visualization environment software to show the electrical field at the head surface (positive potential at red electrodes and negative potential at blue electrodes) created by electrical activity in auditory cortex (dipoles shown as gold arrows seen best in the inflated cortex at (b)). ((c), (d)) Through reciprocity theory, the field pattern of the cortical sources can be used to estimate the optimal electrodes (sources: red, sinks: blue) to impress electrical current on the auditory cortex for neuromodulation. Here, the impressed cortical current density is imaged with a hot metal palette (gold, orange, and yellow), thresholded at 50% of maximum.

CONCLUSION

BrainK is a software package that implements certain tissue segmentation and surface extraction algorithms that are useful in scientific and medical image processing. It also applies certain algorithms, such as skull image registration and cortical surface dipole tessellation, which are specifically important to model the electrical properties of the human head. Accurate electrical head models are proving important for improving the source analysis of human electrophysiological activity with dense array technologies and for optimizing transcranial current delivery to modulate brain activity.

Competing Interests

Kai Li and Don M. Tucker are employees of EGI, a medical device company that is developing medical software from the methods described here. EGI and the University of Oregon hold a US patent (no. 8,478,011) on the relative thresholding method. Xenophon Papademetris is a paid consultant for EGI.

Acknowledgments

The first author would like to thank his Ph.D. advisors Allen Malony and Dr. Don Tucker for their long and continuing support in the procedure of the research work, Dr. Jasmine Song for her valuable help in revising the paper, and Dr. Phan Luu, Dr. Sergei Turovets, Dr. David Hammond, Chelsea Mattson, Colin Davey, Ryan Jayne, and Kyle Morgan for their constructive comments, advice, and requests provided in the process of using and testing the BrainK software. Xenophon Papademetris worked on this project as a paid consultant of Electrical Geodesics, Inc. The intellectual property of this software is protected by US Pat. no. 8,478,011.

REFERENCES

1. D. M. Tucker, "Spatial sampling of head electrical fields: the geodesic sensor net," Electroencephalography and Clinical Neurophysiology, vol. 87, no. 3, pp. 154–163, 1993.
2. J. Song, D. M. Tucker, T. Gilbert et al., "Methods for examining electrophysiological coherence in epileptic networks," Frontiers in Neurology, vol. 4, article 55, 2013.
3. D. L. Schomer and F. L. da Silva, Eds., Niedermeyer's Electroencephalography: Basic Principles, Clinical Applications, and Related Fields, Lippincott Williams & Wilkins, 6th edition, 2011.

4. G. S. Russell, K. J. Eriksen, P. Poolman, P. Luu, and D. M. Tucker, "Geodesic photogrammetry for localizing sensor positions in dense-array EEG," Clinical Neurophysiology, vol. 116, no. 5, pp. 1130–1140, 2005.
5. S. Gonçalves, J. C. de Munck, J. P. Verbunt, R. M. Heethaar, and F. H. da Silva, "In vivo measurement of the brain and skull resistivities using an EIT-based method and the combined analysis of SEF/SEP data," IEEE Transactions on Biomedical Engineering, vol. 50, no. 9, pp. 1124–1128, 2003.
6. Ü. Aydin, J. Vorwerk, P. Küpper et al., "Combining EEG and MEG for the reconstruction of epileptic activity using a calibrated realistic volume conductor model," PLoS ONE, vol. 9, no. 3, Article ID e93154, 2014.
7. M. Dannhauer, R. MacLeod, S. Guler et al., "Comparison of patch electrodes and high density electrode arrays regarding volume conduction in tdcs using a realistic head model," in Proceedings of the 19th Annual Meeting of the Organization for Human Brain Mapping (OHBM ‹13), Seattle, Wash, USA, June 2013.
8. S. Guler, M. Dannhauer, R. MacLeod et al., "Dense electrode array current optimization for targeting and directionality of tDCS: studies in a realistic head model," in Proceedings of the Organization for Human Brain Mapping Meeting (OHBM ‹13), Seattle, Wash, USA, June 2013.
9. F. Klauschen, A. Goldman, V. Barra, A. Meyer-Lindenberg, and A. Lundervold, "Evaluation of automated brain MR image segmentation and volumetry methods," Human Brain Mapping, vol. 30, no. 4, pp. 1310–1327, 2009.
10. S. M. Smith, M. Jenkinson, M. W. Woolrich et al., "Advances in functional and structural MR image analysis and implementation as FSL," NeuroImage, vol. 23, pp. S208–S219, 2004.
11. A. M. Dale, B. Fischl, and M. I. Sereno, "Cortical surface-based analysis: I. Segmentation and surface reconstruction," Neuroimage, vol. 9, no. 2, pp. 179–194, 1999.
12. B. Fischl, M. I. Sereno, and A. M. Dale, "Cortical surface-based analysis. II: inflation, flattening, and a surface-based coordinate system," NeuroImage, vol. 9, no. 2, pp. 195–207, 1999.

13. J. Ashburner and K. J. Friston, "Unified segmentation," NeuroImage, vol. 26, no. 3, pp. 839–851, 2005.
14. S. Turovets, A. Salman, A. Malony, P. Poolman, C. Davey, and D. M. Tucker, "Anatomically constrained conductivity estimation of the human head tissues in vivo: computational procedure and preliminary experiments," in Proceedings of the 7th Conference on Biomedical Applications of Electrical Impedance Tomography, Seoul, South Korea, 2006.
15. K. Li, A. D. Malony, and D. M. Tucker, "Automatic brain MR image segmentation by relative thresholding and morphological image analysis," in Proceedings of the 1st International Conference on Computer Vision Theory and Applications (VISAPP ‹06), pp. 354–364, Setubal, Portugal, 2006.
16. T. H. Cormen, C. E. Leiserson, R. L. Rivest, and C. Stein, Introduction to Algorithms, MIT Press and McGraw-Hill, 2nd edition, 2001.
17. J. Sierpowska, M. A. Hakulinen, J. Töyräs et al., "Interrelationships between electrical properties and microstructure of human trabecular bone," Physics in Medicine and Biology, vol. 51, no. 20, pp. 5289–5303, 2006.
18. N. Otsu, "A threshold selection method from gray-level histograms," IEEE Transactions on Systems, Man, and Cybernetics, vol. 9, no. 1, pp. 62–66, 1979.
19. A. P. Zijdenbos and B. M. Dawant, "Brain segmentation and white matter lesion detection in MR images," Critical Reviews in Biomedical Engineering, vol. 22, no. 5-6, pp. 401–465, 1994.
20. G. Bertrand, "Simple points, topological numbers and geodesic neighborhoods in cubic grids," Pattern Recognition Letters, vol. 15, no. 10, pp. 1003–1011, 1994.
21. K. Li, A. D. Malony, and D. M. Tucker, "A multiscale morphological approach to topology correction of cortical surfaces," in Medical Imaging and Augmented Reality: Third International Workshop, Shanghai, China, August 17–18, 2006 Proceedings, vol. 4091 of Lecture Notes in Computer Science, pp. 52–59, Springer, Berlin, Germany, 2006.
22. J.-D. Boissonnat and M. Yvinec, Algorithmic Geometry, Cambridge University Press, 1998.View at MathSciNet

23. M. H. Davis, A. Khotanzad, D. P. Flamig, and S. E. Harms, "A physics-based coordinate transformation for 3-D image matching," IEEE Transactions on Medical Imaging, vol. 16, no. 3, pp. 317–328, 1997.

24. B. K. Natarajan, "On generating topologically consistent isosurfaces from uniform samples," The Visual Computer, vol. 11, no. 1, pp. 52–62, 1994.

25. B. L. Chamberlain, "Graph partitioning algorithms for distributing workloads of parallel computations," Tech. Rep. TR-98-10-03, Department of Computer Science & Engineering, University of Washington, 1998.

26. B. Hendrickson and R. Leland, "The Chaco users guide: version 2.0," Tech. Rep. SAND 94-2692, Sandia National Laboratories, Albuquerque, NM, USA, 1994.

27. J. Talairach and P. Tournoux, Co-Planar Stereotaxic Atlas of the Human Brain, Thieme, New York, NY, USA, 1988.

28. J. L. Lancaster, M. G. Woldorff, L. M. Parsons et al., "Automated Talairach Atlas labels for functional brain mapping," Human Brain Mapping, vol. 10, no. 3, pp. 120–131, 2000.

29. B. Fischl, D. Salat, E. Busa et al., "Whole brain segmentation: automated labeling of neuroanatomical structures in the human brain," Neuron, vol. 33, no. 3, pp. 341–355, 2002.

30. Freesurfer, https://surfer.nmr.mgh.harvard.edu.

31. SPM, http://www.fil.ion.ucl.ac.uk/spm.

32. J. Ashburner and K. J. Friston, "Unified segmentation," NeuroImage, vol. 26, no. 3, pp. 839–851, 2005.

33. FSL, http://www.fmrib.ox.ac.uk/fsl.

34. S. M. Smith, "Fast robust automated brain extraction," Human Brain Mapping, vol. 17, no. 3, pp. 143–155, 2002.

35. Y. Zhang, M. Brady, and S. Smith, "Segmentation of brain MR images through a hidden Markov random field model and the expectation-maximization algorithm," IEEE Transactions on Medical Imaging, vol. 20, no. 1, pp. 45–57, 2001.

36. Y. Cointepas, J.-F. Mangin, L. Garnero, J.-B. Poline, and H. Benali, "BrainVISA: software platform for visualization and analysis of

multi-modality brain data," in Proceedings of the 7th Human Brain Mapping Conference, p. S98, Brighton, UK, 2001.

37. BrainVisa, http://brainvisa.info.
38. J.-F. Mangin, "Entropy minimization for automatic correction of intensity nonuniformity," in Proceedings of the IEEE Workshop on Mathematical Methods in Biomedical Image Analysis (MMBIA ‹00), pp. 162–169, Hilton Head Island, SC, USA, June 2000.
39. J.-F. Mangin, O. Coulon, and V. Frouin, "Robust brain segmentation using histogram scale-space analysis and mathematical morphology," in Proceedings of the Medical Image Computing and Computer-Assisted Intervention (MICCAI ‹98), W. M. Wells, A. Colchester, and S. Delp, Eds., pp. 1230–1241, MIT, Boston, Mass, USA, 1998.
40. J.-F. Mangin, J. Régis, and V. Frouin, "Shape bottlenecks and conservative flow systems," in Proceedings of the IEEE/SIAM Workshop on Mathematical Methods in Biomedical Image Analysis (MMBIA ‹96), pp. 319–328, San Francisco, Calif, USA, 1996.
41. J.-F. Mangin, O. Coulon, and V. Frouin, "Robust brain segmentation using histogram scale-space analysis and mathematical morphology," in Medical Image Computing and Computer-Assisted Interventation—MICCAI ‹98, W. M. Wells, A. Colchester, and S. Delp, Eds., vol. 1496 of Lecture Notes in Computer Science, pp. 1230–1241, 1998.
42. C. Cocosco, V. Kollokian, R.-S. Kwan, and A. Evans, "BrainWeb: online interface to a 3D MRI simulated brain database," NeuroImage, vol. 5, no. 4, p. S425, 1997.
43. BrainWeb, http://www.bic.mni.mcgill.ca/brainweb.
44. CMA: The Center for Morphometric Analysis at Massachusetts General Hospital, http://www.cma.mgh.harvard.edu/ibsr/.
45. NRU: The Neurobiology Research Unit, http://nru.dk.
46. D. W. Shattuck, S. R. Sandor-Leahy, K. A. Schaper, D. A. Rottenberg, and R. M. Leahy, "Magnetic resonance image tissue classification using a partial volume model," NeuroImage, vol. 13, no. 5, pp. 856–876, 2001.
47. K. van Leemput, F. Maes, D. Vandermeulen, and P. Suetens, "Automated model-based tissue classification of MR images of the

brain," IEEE Transactions on Medical Imaging, vol. 18, no. 10, pp. 897–908, 1999.

48. K. Li, Neuroanatomical segmentation in MRI exploiting a priori knowledge [Ph.D. thesis], University of Oregon, 2007.
49. A. Salman, S. Turovets, A. Malony, J. Eriksen, and D. M. Tucker, "Computational modeling of human head conductivity," in Proceedings of the 5th International Conference on Computational Science (ICCS ‹05), Atlanta, Ga, USA, May 2005.
50. A. Salman, S. Turovets, A. Malony et al., "Noninvasive conductivity extraction for high-resolution eeg source localization," Advances in Clinical Neuroscience and Rchabilitation, vol. 26, no. 1, pp. 27–28, 2005.

SECTION IV: NEURAL NETWORK APPLICATIONS

CHAPTER 10

QUANTUM-INSPIRED NEURAL NETWORKS WITH APPLICATION

Jianping Li

School of Computer and Information Technology, Northeast Petroleum University, Daqing, China

ABSTRACT

In this paper, a novel neural network is proposed based on quantum rotation gate and controlled- NOT gate. Both the input layer and the hide layer are quantum-inspired neurons. The input is given by qubits, and the output is the probability of qubit in the state $|1\rangle$. By employing the gradient descent method, a training algorithm is introduced. The experimental results show that this model is superior to the common BP networks.

Keywords: Quantum Computing, Neural Networks, Algorithm Design

INTRODUCTION

In the eighties of the twentieth century, Benioff and Feynman proposed the concept of quantum computation, and then P.W. Shor gave the first quantum algorithm of very large integer factorization [1] in 1994 and L.K. Grover proposed a quantum algorithm which searches a marked state in an unordered list [2] in 1996. Quantum computation has been widely paid attention and become a challenging research line. Fuzzy logic, evolution calculation, and neural networks are regarded as the most promising three important aspects in the artificial intelligence field, which compose intelligence calculation (soft calculation) and have much comparability with quantum computation. Therefore, the combination of them would bring promising research in theory.

Some research results in the nineties of the twentieth century show that [3] - [5] the information processing of brain may be related to the quantum states; there may exists the effect of quantum mechanics in brain, and the quantum system has the same dynamics features as the biology neural networks. Fatt and Katz proved that in biology neuron, the neurotransmitter is released from the nerve terminals by the form of multi-molecule and small units, and they call this kind of small units quantum. Each of small units (quantum) describes the minimal unit that the neurotransmitter is released. In general, although the number of quantum in each synaptic response of neurons may be different, the number of molecules in each quantum is the same. Therefore, the combination of quantum computation and ANN may preferably simulate the information processing of brain.

However, as yet, there is little understanding of the essential components of artificial neural networks based on quantum theoretical concepts and techniques. The basal model and theory of quantum neural networks are in research. At present, there is not a set of perfect theory to direct the construction of model. Since Kak firstly proposed the concept of quantum neural computation [6] in 1995, the quantum neural networks have been paid attention, and many novel ideas and elementary model have been proposed. In 1997, N. B. Karayiannis et al. proposed the model of quantum neural networks with multilevel hidden neurons based on the superposition of quantum states in quantum theory. ZHU Da-qi et al. applied this model to fault diagnosis for photovoltaic radar electronic equipment, and acquired the satisfied results [7] . In 2000, Matsui et al. proposed a quantum neural networks model based on the single bit rotation gate and the 2-bit controlled-NOT gate, presented the algorithm of model, and investigated its performance on solving the 4-bit parity check problem and the function approximation problem [8] . In this model, the input is based on quantum bits (qubits). However, in many actual problems, the system input is the real vector in Hilbert space, and the method is not presented to transform the real to the quantum state in Ref. [8] . For the algorithm of model, the only two iterative equations in complex number are given based on the gradient descent algorithm. Neither the gradient computation formula nor the input-output relationship of networks is presented. Therefore, it is not easy for the reader to simulate the algorithm and apply this model to actual problem.

In this paper, based on the universality of the single qubit rotation gate and the two-qubit controlled-NOT gate, a quantum neuron model and three layers quantum back-propagation neural networks (QBP) model are proposed. The learning algorithm in Ref. [8] is improved. A transform method from the real to the qubit is proposed. On the basis of the probability amplitude vector of qubits, the input-output relationship of the model is presented. To facilitate the practical application, the learning algorithm is deducted in detail and the implementing scheme is given. The continuity of this model is theoretically proved. Two application examples are designed, and the simulation results show that this model and algorithm are evidently superior to the conventional back-propagation networks (CBP) in three aspects: convergence speed, convergence rate, and robustness.

THE QUBIT AND QUANTUM GATES

The Qubit

In the quantum computation, the term "qubit" is introduced as the counterpart of the "bit" in the conventional computation to describe the state of quantum computation. In quantum computation system two quantum physical states labeled as $|0\rangle$ and $|1\rangle$ express 1 bit information. $|0\rangle$ corresponds to the bit 0 of classical computation, while the $|1\rangle$ to the bit 1. The qubit state $|\phi\rangle$ maintains a coherent superposition of states $|0\rangle$ and $|1\rangle$

$$|\phi\rangle = \alpha|0\rangle + \beta|1\rangle \qquad (1)$$

where α and β are complex numbers called probability amplitude. That is, the qubit state $|\phi\rangle$ collapses into either $|0\rangle$ state with probability $|\alpha|^2$, or $|1\rangle$ state with probability $|\beta|^2$, and

$$|\alpha|^2 + |\beta|^2 = 1 \qquad (2)$$

Therefore, the qubit can also be described by the probability amplitudes as $[\alpha, \beta]^T$.

The Quantum Gate

The definition of single qubit rotation gate is described as follows

$$R(\theta) = \begin{bmatrix} \cos\theta & -\sin\theta \\ \sin\theta & \cos\theta \end{bmatrix} \qquad (3)$$

Let the quantum state $|\phi\rangle = \begin{bmatrix} \cos\theta_0 \\ \sin\theta_0 \end{bmatrix}$. Then the $|\phi\rangle$ can be evolved by R(θ) as follows $R(\theta)|\phi\rangle = \begin{bmatrix} \cos(\theta_0 + \theta) \\ \sin(\theta_0 + \theta) \end{bmatrix}$. It is obvious that the R(θ) performs the phase rotation.

The effect of the quantum NOT gate is exchange the two probability amplitudes of a qubit, and its definition is described as follows

$$U = \begin{bmatrix} 0 & 1 \\ 1 & 0 \end{bmatrix} \qquad (4)$$

Using the method proposed in Ref. [8] , the controlled-NON gate can be realized by Equation (5)

$$C(k) = \begin{bmatrix} \cos\left(\frac{\pi}{2}k - 2\theta_0\right) & -\sin\left(\frac{\pi}{2}k - 2\theta_0\right) \\ \sin\left(\frac{\pi}{2}k - 2\theta_0\right) & \cos\left(\frac{\pi}{2}k - 2\theta_0\right) \end{bmatrix} \qquad (5)$$

According to the different value of the controlled parameter k, the controlled effect includes three cases as follows

(1) If k=1, then $C(k)|\phi\rangle = \begin{bmatrix} \sin\theta_0 \\ \cos\theta_0 \end{bmatrix}$, which corresponds to reversal rotation;

(2) If k=0, then $C(k)|\phi\rangle = \begin{bmatrix} \cos\theta_0 \\ -\sin\theta_0 \end{bmatrix}$, which corresponds to not rotation. In this case, the phase of the pro-

bability amplitude of quantum state $|1\rangle$ is reversed. However, its observed probability is invariant so that we are able to regard this case as no-rotation;

(3) If 0<k<1, then

$$C(k)|\phi\rangle = \begin{bmatrix} \cos\left(\frac{\pi}{2}k - \theta_0\right) \\ \sin\left(\frac{\pi}{2}k - \theta_0\right) \end{bmatrix} \quad (6)$$

THE QUANTUM-INSPIRED NEURAL NETWORKS MODEL

The Quantum-Inspired Neuron

On the basis of the universal quantum gates, the quantum neuron model proposed in this paper includes five parts: input, phase rotation, aggregation, reverse rotation and output, where the input is described by qubits, the output is given by the probability of the state in which $|1\rangle$ is observed, the phase rotation and the reverse rotation are performed by quantum rotation gate $R(\theta_i)$ and controlled-NOT gate $U(\gamma)$, respectively. The quantum neuron model is shown in Figure 1.

Where the definition of $R(\theta_i)$ is same as Equation (3). The $U(\gamma)$ is defined as follows

$$U(\gamma) = C(f(\gamma)) \quad (7)$$

where f is a sigmoid function, and the definition of $C(\cdot)$ refers to Equation (5). Let $|x_i\rangle = [\cos t_i \quad \sin t_i]^{\mathrm{T}}$, and $\theta = \arg\left(\sum_{i=1}^{n} R(\theta_i)|x_i\rangle\right) = \arctan\left(\sum_{i=1}^{n}\sin(t_i+\theta_i) \Big/ \sum_{i=1}^{n}\cos(t_i+\theta_i)\right)$. The result of aggregation is defined as follows

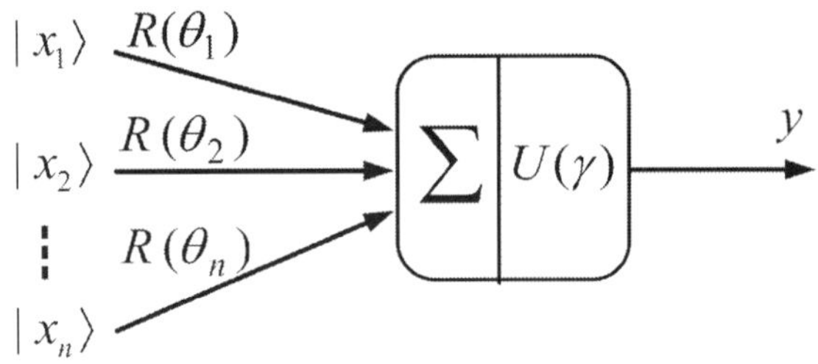

Figure 1.The quantum-inspired neuron model.

$$\sum_{i=1}^{n} R(\theta_i)|x_i\rangle = [\cos\theta \quad \sin\theta]^{T} \tag{8}$$

The result of reverse rotation is given by

$$U(\gamma)\sum_{i=1}^{n} R(\theta_i)|x_i\rangle = \left[\cos\left(\frac{\pi}{2}f(\gamma)-\theta\right) \quad \sin\left(\frac{\pi}{2}f(\gamma)-\theta\right)\right]^{T} \tag{9}$$

The input-output relationship of quantum neuron is described as follows

$$y = \sin\left(\frac{\pi}{2}f(\gamma)-\theta\right) = \sin\left(\frac{\pi}{2}f(\gamma)-\arg\left(\sum_{i=1}^{n} R(\theta_i)|x_i\rangle\right)\right) \tag{10}$$

The Quantum-Inspired Neural Networks

The quantum neural networks are composed of some quantum neurons and conventional neurons according to the definite linking rule. The three layers feed-forward QBP model proposed in this paper is shown in Figure 2, where the input layer and the hidden layer has n, p quantum neurons, respectively, and the output layer has m conventional neurons.

Suppose $|x_i\rangle$ are the networks input, h_j are the hide layer output, y_k are the networks output, $R(\theta_{ij})$ are the quantum rotation gates to update qubits in the hide layer, w_{jk} are the link-weights between the hide layer and the output layer, C(0)and $U(\alpha_j)$ are the controlled-NOT gates that are regarded as the transformation function of quantum neurons in the input layer and hide layer, respectively. The input-output relationship of each layer is described as follows

$$h_j = \sin\left(\frac{\pi}{2}f(\alpha_j)-\arg\left(\sum_{i=1}^{n} R(\theta_{ij})|x_i\rangle\right)\right) \tag{11}$$

$$y_k = g\left(\sum_{j=1}^{p} w_{jk} h_j\right) = g\left(\sum_{j=1}^{p} w_{jk} \sin\left(\frac{\pi}{2} f(\alpha_j) - \arg\left(\sum_{i=1}^{n} R(\theta_{ij}) |x_i\rangle\right)\right)\right) \quad (12)$$

Where i=1, 2, ...n; j=1, 2, ...p; k=1, 2, ...m.

Training Algorithm

For the training samples described in the n-dimension Euclid-space $\overline{X} = (\overline{x}_1, \overline{x}_2, \cdots, \overline{x}_n)^{\mathrm{T}}$, the transform formula to realize quantum state description of the training samples is defined as follows

$$|X\rangle = [|x_1\rangle, |x_2\rangle, \cdots, |x_n\rangle]^{\mathrm{T}} \quad (13)$$

Where

$$|x_i\rangle = \cos\left(\frac{2\pi}{1+\exp(-\overline{x}_i)}\right)|0\rangle + \sin\left(\frac{2\pi}{1+\exp(-\overline{x}_i)}\right)|1\rangle = \left[\cos\left(\frac{2\pi}{1+\exp(-\overline{x}_i)}\right) \quad \sin\left(\frac{2\pi}{1+\exp(-\overline{x}_i)}\right)\right]^{\mathrm{T}}$$

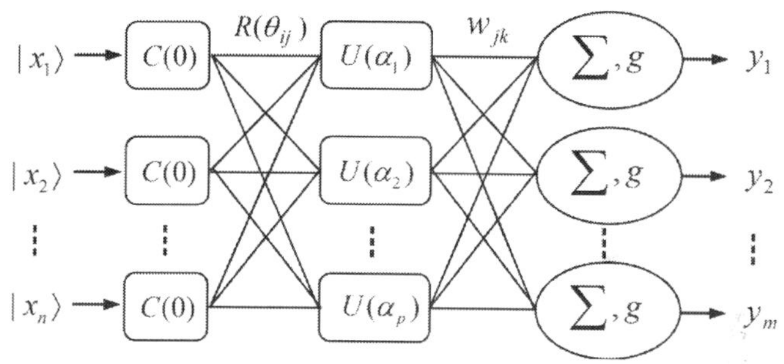

Figure 2. Three layers quantum-inspired networks model.

In the QBP described by Figure 2, there are three groups of parameters, the rotation parameter θ_{ij}, the reversal parameter α_j and the link-weights w_{jk}, that need updating. The error function is defined as follows

$$E = \frac{1}{2}\left(\tilde{Y} - Y\right)^2 \quad (14)$$

where $\tilde{Y}$ and Y are the normalized desired output and the practical output, respectively.

Let $|x_i\rangle = \begin{bmatrix}\cos\varphi_i \\ \sin\varphi_i\end{bmatrix}$, $\beta_j = \arctan\left(\frac{\sum_{i=1}^{n}\sin(\varphi_i + \theta_{ij})}{\sum_{i=1}^{n}\cos(\varphi_i + \theta_{ij})}\right)$, Equation (12) can be rewritten as

$$y_k = g\left(\sum_{j=1}^{p} w_{jk} \sin\left(\frac{\pi}{2} f\left(\alpha_j\right) - \beta_j\right)\right) \tag{15}$$

Let

$$S_j = \frac{\sum_{i=1}^{n} \sin\left(\varphi_i + \theta_{ij}\right)}{\sum_{i=1}^{n} \cos\left(\varphi_i + \theta_{ij}\right)},$$

$$S_{j1} = \sum_{i=1}^{n} \cos\left(\varphi_i + \theta_{ij}\right),$$

$$T_j = \frac{\cos\left(\varphi_i + \theta_{ij}\right) S_{j1} + \sin^2\left(\varphi_i + \theta_{ij}\right)}{S_{j1}^2}.$$

According to the gradient descent algorithm,

$$\Delta\theta_{ij} = -\frac{\partial E}{\partial \theta_{ij}} = -\sum_{k=1}^{m} (\tilde{y}_k - y_k) g' w_{jk} \cos\left(\frac{\pi}{2} f\left(\alpha_j\right) - \beta_j\right) \frac{T_j}{1 + S_j^2} \tag{16}$$

$$\Delta\alpha_j = -\frac{\partial E}{\partial \alpha_j} = \frac{\pi}{2} \sum_{k=1}^{m} (\tilde{y}_k - y_k) g' w_{jk} \cos\left(\frac{\pi}{2} f\left(\alpha_j\right) - \beta_j\right) f' \tag{17}$$

$$\Delta w_{jk} = -\frac{\partial E}{\partial w_{jk}} = (\tilde{y}_k - y_k) g' \sin\left(\frac{\pi}{2} f\left(\alpha_j\right) - \beta_j\right) \tag{18}$$

The parameter updating rules are as follows

$$\theta_{ij}(t+1) = \theta_{ij}(t) + \eta \Delta\theta_{ij}(t) \tag{19}$$

$$\alpha_j(t+1) = \alpha_j(t) + \eta \Delta\alpha_j(t) \tag{20}$$

$$w_{jk}(t+1) = w_{jk}(t) + \eta \Delta w_{jk}(t) \tag{21}$$

where η is the learning coefficient.

SIMULATION COMPARISONS

To testify the validity of QBP, an actual example is designed and the QBP is compared with the CBP in this section. To make comparison equitable, the QBP adopts the same structure and parameters as CBP in the simulations.

The XOR problem has been the typical example of the neural networks learning algorithm research. This problem is the simplest example for nonlinear classification problem, which the corresponding optimization surface is irregular, and exists some local minimum. In this simulation, The XOR problem is generalized from four points to sixteen points. The point set is shown in Figure 3.

Firstly, we investigate how to change for the convergence rate when the learning coefficient changes. The networks structure is set to 2-10-1, the restriction error is 0.1, and the restriction iteration steps are 2000. The learning coefficient is from $\{0.1, 0.2, \cdot\cdot\cdot, 1.0\}$. This example is simulated 100 times for each learning coefficient by the QBP and the CBP, respectively. When the learning coefficient changes, the maximum of convergence rate of QBP is 87% and the minimum is 51%. However, the convergence rate of CBP changes in a large range when the learning coefficient changes. When the learning coefficient is less than 0.2, the convergence rate of CBP is 0%, and when the learning coefficient is more than 0.2, the maximum of convergence rate of CBP is only 59%. The comparison result is shown in Figure 4.

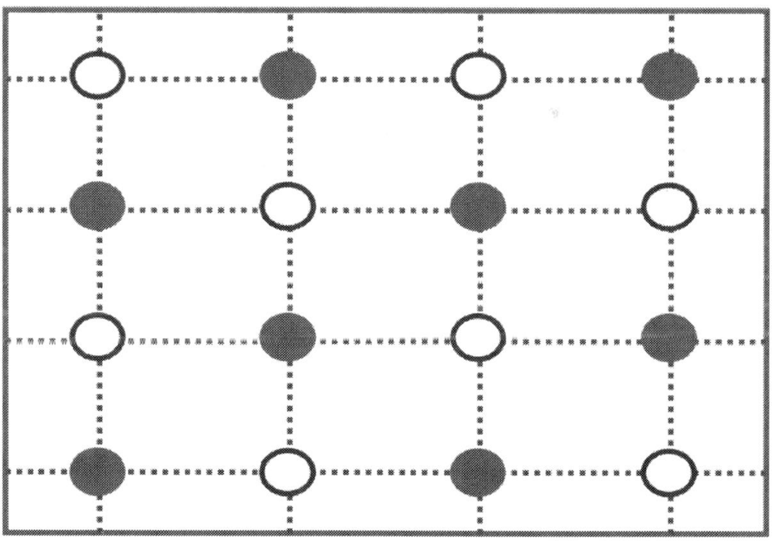

Figure 3. Sixteen points distribution on the plane.

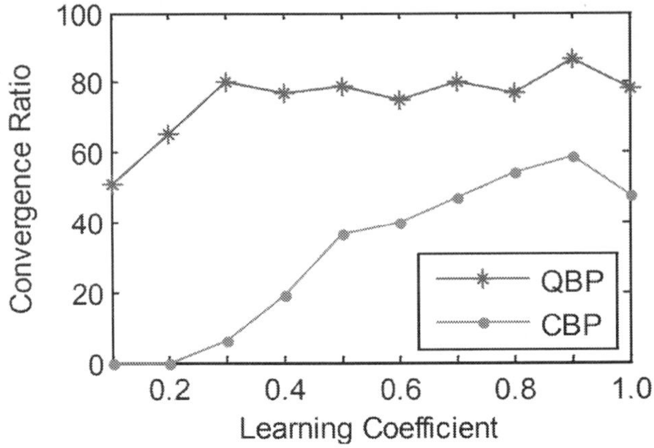

Figure 4. The relation of the convergence rate and η.

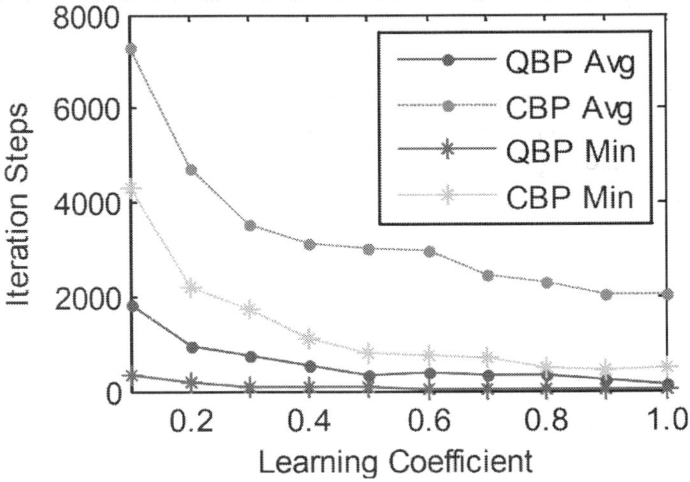

Figure 5. The relation of the iteration steps and η.

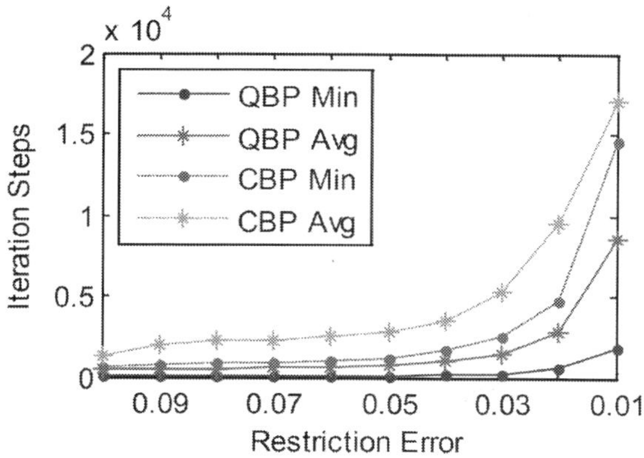

Figure 6. The relation between the iteration steps and the restriction error.

The comparison result shows that the QBP is evidently superior to the CBP in both the convergence rate and the robustness.

Secondly, we investigate how to change for the iteration steps when the learning coefficient changes. The networks structure is set to 2-10-1 , the restriction error is 0.1, and the restriction iteration steps are 10,000. The learning coefficient is still from $\{0.1, 0.2, \cdot \cdot \cdot, 1.0\}$. This example is simulated 100 times for each learning coefficient by the QBP and the CBP, respectively. When the learning coefficient changes, for the average iteration steps of QBP, the maximum is 1838 steps, and the minimum is 151 steps; for the minimum iteration steps of QBP, the maximum is 336 steps, and the minimum is 27 steps. However, for the average iteration steps of CBP, the maximum is 7308 steps, and the minimum is 2021 steps; for the minimum iteration steps of CBP, the maximum is 4314 steps, and the minimum is 463 steps. The comparison result is shown in Figure 5.

The comparison result shows that the QBP is evidently superior to the CBP in both the iteration steps and its fluctuation range when the learning coefficient changes.

Finally, we investigate how to change for the iteration steps when the restriction error changes. The networks structure is set to 2-10-1 . It is known from Figure 4 and Figure 5 that the performance of QBP and CBP is the best when the learning coefficient is set 0.9. Therefore, the learning coefficient is set 0.9. The restriction error is from {0.10, 0.09,

· · · , 0.01}. This example is simulated 100 times for each restriction error by the QBP and the CBP, respectively. The comparison result is shown in Figure 6.

When the restriction error changes, for the average iteration steps of QBP, the maximum is 8575 steps, and the minimum is 431 steps; for the minimum iteration steps of QBP, the maximum is 1816 steps, and the minimum is 30 steps. However, for the average iteration steps of CBP, the maximum is 17,100 steps, and the minimum is 1304 steps; for the minimum iteration steps of CBP, the maximum is 14,525 steps, and the minimum is 659 steps. Hence, when the restriction error changes, the iteration steps and its fluctuation range of QBP is far less than that of CBP, which takes on the preferable robustness.

CONCLUSION

On the basis of the qubits and the universal quantum gates, a quantum BP neural networks model is proposed; the learning algorithm of this model is designed; and the continuity of this model is proved. The simulation result shows that this model and algorithm are superior to conventional BP networks in three aspects: the convergence speed, convergence rate, and robustness, by an actual example of pattern recognition and function approximation.

REFERENCES

1. Shor, P.W. (1994) Algorithms for Quantum Computation: Discrete Logarithms and Factoring. 35th Annual Symposium on Foundations of Computer Science, New Mexico, 20-22 November 1994, 124-134. http://dx.doi.org/10.1109/SFCS.1994.365700
2. Grover, L.K. (1996) A Fast Quantum Mechanical Algorithm for Database Search. Proceedings of the 28th Annual ACM Symposium on the Theory of Computing, Pennsylvania, 212-221. http://dx.doi.org/10.1145/237814.237866
3. Perus, M. (1996) Neuro-Quantum Parallelism in Brain-Mind and Computers. Informatics, 20, 173-183.
4. Adenilton, J., Wilson, R. and Teresa, B. (2012) Classical and Superposed Learning for Quantum Weightless Neural Network.

Neurocomputing, 75, 52-60. http://dx.doi.org/10.1016/j.neucom.2011.03.055

5. Israel, G.C., Angel, G.C. and Belen, R.M. (2011) Dealing with Limited Data in Ballistic Impact Scenarios: An Empirical Comparison of Different Neural Network Approaches. Application Intelligence, 35, 89-109. http://dx.doi.org/10.1007/s10489-009-0205-8
6. Kak, S. (1995) On Quantum Neural Computing. Information Sciences, 19, 143-160. http://dx.doi.org/10.1016/0020-0255(94)00095-S
7. Zhu, D.Q. and Sang, Q.B. (2006) A Fault Diagnosis Algorithm for the Photovoltaic Radar Electronic Equipment Based on Quantum Neural Networks. Acta Electronica Sinica, 34, 573-576.
8. Matsui, N., Kouda, N. and Nishimura, H. (2000) Neural Networks Based on QBP and Its Performance. Proceedings of the IEEE-INNS-ENNS International Joint Conference on Neural Networks, 3, 247-252. http://dx.doi.org/10.1109/ijcnn.2000.861311

CHAPTER 11

TRAINING FEEDFORWARD NEURAL NETWORKS USING SYMBIOTIC ORGANISMS SEARCH ALGORITHM

Haizhou Wu,[1] Yongquan Zhou,[1,2] Qifang Luo,[1,2] and Mohamed Abdel Basset[3]

[1]College of Information Science and Engineering, Guangxi University for Nationalities, Nanning 530006, China

[2]Key Laboratory of Guangxi High Schools Complex System and Computational Intelligence, Nanning 530006, China

[3]Faculty of Computers and Informatics, Zagazig University, Zagazig, Egypt

ABSTRACT

Symbiotic organisms search (SOS) is a new robust and powerful metaheuristic algorithm, which stimulates the symbiotic interaction strategies adopted by organisms to survive and propagate in the ecosystem. In the supervised learning area, it is a challenging task to present a satisfactory and efficient training algorithm for feedforward neural networks (FNNs). In this paper, SOS is employed as a new method for training FNNs. To investigate the performance of the aforementioned method, eight different datasets selected from the UCI machine learning repository are employed for experiment and the results are compared among seven metaheuristic algorithms. The results show that SOS performs better than other algorithms for training FNNs in terms of converging speed. It is also proven that an FNN trained by the method of SOS has better accuracy than most algorithms compared.

INTRODUCTION

Artificial neural networks (ANNs) [1] are mathematical models and have been widely utilized for modeling complex nonlinear processes. As one of the powerful tools, ANNs have been employed in various fields, like time series prediction [2], classification [3, 4], pattern recognition [5–7], system identification and control [8], function approximation [9], signal processing [10], and so on [11, 12].

There are different types of ANNs proposed in the literature: feedforward neural networks (FNNs) [13], Kohonen self-organizing network [14], radial basis function (RBF) [15–17], recurrent neural network [18], and spiking neural networks [19]. In fact, feedforward neural networks are the most popular neural networks in practical applications. Training process is one of the most important aspects for neural networks. In this process, the goal is to achieve the minimum cost function defined as a mean squared error (MSE) or a sum of squared error (SSE) by the means of finding the best combination of connection weights and biases. In general, training algorithms can be classified into two groups: gradient-based algorithms versus stochastic search algorithms. The most widely applied gradient-based training algorithms are backpropagation (BP) algorithm [20] and its variants [21]. However, in complex nonlinear problems, these two algorithms suffer from some shortcomings, such as highly depending on the initial solution, which subsequently impact on the convergence of the algorithm and easily get trapped into local optima. On the other hand, stochastic search methods like metaheuristic algorithms were proposed by researchers as alternatives to gradient-based methods for training FNNs. Metaheuristic algorithms are proved to be more efficient in escaping from local minima for optimization problems.

Various metaheuristic optimization methods have been used to train FNNs. GA, inspired by Darwinians' theory of evolution and natural selection [22], is one of the earliest methods for training FNNs that proposed by Montana and Davis [23]. The results indicate that GA is able to outperform BP when solving real and challenging problems. Shaw and Kinsner presented a method called chaotic simulated annealing [24, 25], which is superior in escaping from local optima for training multilayer FNNs. Zhang et al. proposed a hybrid particle swarm optimization-backpropagation algorithm for feedforward neural network training. In their research, a heuristic way was adopted to give a transition from particle swarm search to gradient descending search [26]. In 2012, Mirjalili et al. proposed a hybrid particle swarm optimization (PSO) and gravitational search algorithm (GSA) [27] to train FNNs [28]. The results showed that PSOGSA outperforms both PSO and GSA in terms of converging speed and avoiding local optima and has better accuracy than GSA in the training process. In 2014, a new metaheuristic algorithm called centripetal accelerated particle swarm optimization (CAPSO) was employed by Beheshti et al. to evolve the accuracy in training ANN [29]. Recently, several other metaheuristic algorithms are applied on the research of NNs. In 2014, Pereira et al. introduced social-spider optimization (SSO) to improve the training phase of ANN with multilayer perceptrons and validated the proposed approach in the context of Parkinson's disease recognition [30]. Uzlu et al. applied the ANN model with the teaching-learning-based optimization (TLBO) algorithm to estimate energy consumption in Turkey [31]. In 2016, Kowalski and Łukasik invited the krill herd algorithm (KHA) for learning an artificial neural network (ANN), which has been verified for the classification task [32]. In 2016, Faris et al. employed the recently proposed nature-inspired algorithm called multiverse optimizer (MVO) for training the feedforward neural network. The comparative study demonstrates that MVO is very competitive and outperforms other training algorithms in the majority of datasets [33]. Nayak et al. proposed a firefly based higher order neural network for data classification for maintaining fast learning and avoids the exponential increase of processing units [34]. Many other metaheuristic algorithms, like ant colony optimization (ACO) [35, 36],

Cuckoo Search (CS) [37], Artificial Bee Colony (ABC) [38, 39], Charged System Search (CSS) [40], Grey Wolf Optimizer (GWO) [41], Invasive Weed Optimization (IWO) [42], and Biogeography-Based Optimizer (BBO) [43] have been adopted for the research of neural network.

In this paper, a new method of symbiotic organisms search (SOS) is used for training FNNs. Symbiotic organisms search [44], proposed by Cheng and Prayogo in 2014, is a new swarm intelligence algorithm simulating the symbiotic interaction strategies adopted by organisms to survive and propagate in the ecosystem. And the algorithm has been applied to resolve some engineering design problems by scholars. In 2016, Cheng et al. researched on optimizing multiple-resources leveling in multiple projects using discrete symbiotic organisms search [45]. Eki et al. applied SOS to solve the capacitated vehicle routing problem [46]. Prasad and Mukherjee have used SOS for optimal power flow of power system with FACTS devices [47]. Abdullahi et al. proposed SOS-based task scheduling in cloud computing environment [48]. Verma et al. investigated SOS for congestion management in deregulated environment [49]. Time-cost-labor utilization tradeoff problem was solved by Tran et al. using this algorithm [50]. Recently, in 2016, more and more scholars get interested in the research of the SOS algorithm. Yu et al. applied two solution representations to transform SOS into an applicable solution approach for the capacitated vehicle and then apply a local search strategy to improve the solution quality of SOS [51]. Panda and Pani presented hybrid SOS algorithm with adaptive penalty function to solve multiobjective constrained optimization problems [52]. Banerjee and Chattopadhyay presented a novel modified SOS to design an improved three-dimensional turbo code [53]. Das et al. used SOS to determine the optimal size and location of distributed generation (DG) in radial distribution network (RDN) for the reduction of network loss [54]. Dosoglu et al. utilized SOS for economic/emission dispatch problem in power systems [55].

The structure of this paper is organized as follows. Section 2 gives a brief description of feedforward neural network; Section 3 elaborates the symbiotic organisms Search and Section 4 describes the SOS-based trainer and how it can be used for training FNNs in detail. In Section 5, series of comparison experiments are conducted; our conclusion will be given in Section 6.

FEEDFORWARD NEURAL NETWORK

In the artificial neural network, the feedforward neural network (FNN) was the simplest type which consists of a set of processing elements called "neurons" [33]. In this network, the information moves in only one direction, forward, from the input layer, through the hidden layer and to the output layer. There are no cycles or loops in the network. An example of a simple FNN with a single hidden layer is shown in Figure 1. As shown, each neuron computes the sum of the inputs weight at the presence of a bias and passes this sum through an activation function (like sigmoid function) so that the output is obtained. This process can be expressed as (1) and (2).

$$h_j = \sum_{i=1}^{R} \mathrm{iw}_{j,i} x_i + \mathrm{hb}_j, \qquad (1)$$

where $\mathrm{iw}_{j,i}$ is the weight connected between neurons $i = (1, 2, \ldots, R)$ and $j = (1, 2, \ldots, N)$, hb$j$ is a bias in hidden layer, R is the total number of neurons in input layer, and xi is the corresponding input data. Here, the S-shaped curved sigmoid function is used as the activation function, which is shown in

$$f_{(x)} = \frac{1}{1 + e^{-x}} \qquad (2)$$

Therefore, the output of the neuron in hidden layer can be described as in

$$\mathrm{ho}_j = f_j(h_j) = \frac{1}{\left(1 + e^{-h_j}\right)} \qquad (3)$$

In the output layer, the output of the neuron is shown in

$$y_k = f_k\left(\sum_{j=1}^{N} \mathrm{hw}_{k,j} \mathrm{ho}_j + \mathrm{ob}_k\right) \qquad (4)$$

where $\mathrm{hw}_{j,i}$ is the weight connected between neurons $j = (1, 2, \ldots, N)$ and $k = (1, 2, \ldots, S)$, ob_k is a bias in output layer N is the total number of neurons in hidden layer, and S is the total number of neurons in output layer.

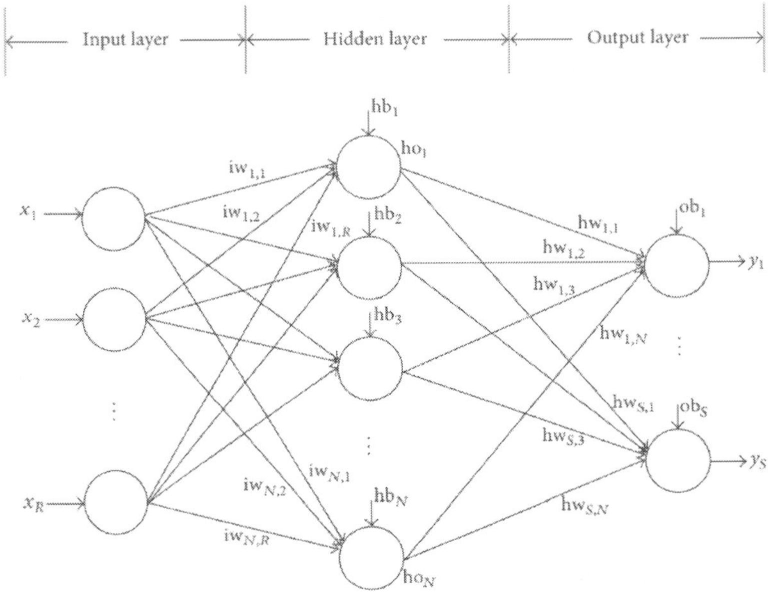

Figure 1: A feedforward network with one hidden layer.

The training process is carried out to adjust the weights and bias until some error criterion is met. Above all, onc problem is to select a proper training algorithm. Also, it is very complex to design the neural network because many elements affect the performance of training, such as the number of neurons in hidden layer, interconnection between neurons and layer, error function, and activation function.

SYMBIOTIC ORGANISMS SEARCH ALGORITHM

Symbiotic organisms search [44] stimulates symbiotic interaction relationship that organisms use to survive in the ecosystem. Three phases, mutualism phase, commensalism phase, and parasitism phase, stimulate the real-world biological interaction between two organisms in ecosystem.

Mutualism Phase

Organisms engage in a mutuality relationship with the goal of increasing mutual survival advantage in the ecosystem. New candidate organisms for X_i and X_j are calculated based on the mutuality symbiosis between organism X_i and X_j, which is modeled in (5) and (6).

$$X_{inew} = X_i + \alpha\left(X_{best} - \text{Mutual\_Vector} * BF_1\right) \quad (5)$$

$$X_{jnew} = X_j + \beta\left(X_{best} - \text{Mutual\_Vector} * BF_2\right) \quad (6)$$

$$\text{Mutual\_Vector} = \frac{\left(X_i + X_j\right)}{2}, \quad (7)$$

where BF_1 and BF_2 are benefit factors that are determined randomly as either 1 or 2. These factors represent partially or fully level of benefit to each organism. X_{best} represents the highest degree of adaptation organism. α and β are random number in [0, 1]. In (7), a vector called "Mutual Vector" represents the relationship characteristic between organisms X_i and X_j.

Commensalism Phase

One organism obtains benefit and does not impact the other in commensalism phase. Organism X_j represents the one that neither benefits nor suffers from the relationship and the new candidate organism of X_i is calculated according to the commensalism symbiosis between organisms X_i and X_j which is modeled in

$$X_{inew} = X_i + \delta\left(X_{best} - X_j\right) \quad (8)$$

where δ represents a random number in [−1, 1]. And X_{best} is the highest degree of adaptation organism.

Parasitism Phase

One organism gains benefit but actively harms the other in the parasitism phase. An artificial parasite called "Parasite Vector" is created in the search space by duplicating organism X_i and then modifying the randomly selected dimensions using a random number. Parasite Vector tries to replace another organism X_j in the ecosystem. According to Darwin's evolution theory, "only the fittest organisms will prevail"; if Parasite Vector is better, it will kill organism X_j and assume its position; else X_j will have immunity from the parasite and the Parasite Vector will no longer be able to live in that ecosystem.

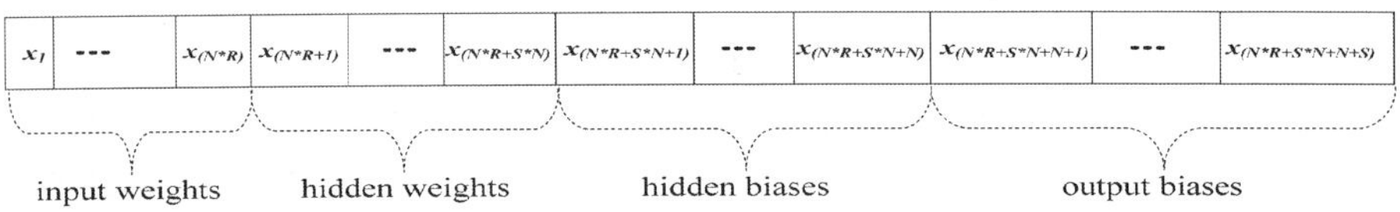

Figure 2: The vector of training parameters.

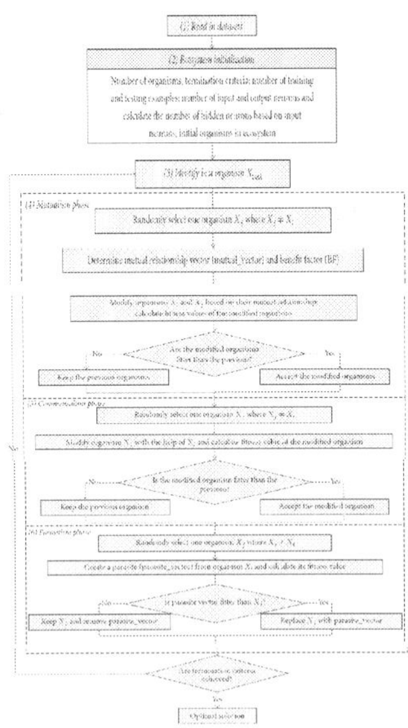

Figure 3: Flowchart of SOS algorithm.

SOS FOR TRAIN FNNS

In this paper, symbiotic organisms search is used as a new method to train

FNNs. The set of weights and bias is simultaneously determined by SOS in order to minimize the overall error of one FNN and its corresponding accuracy by training the network. This means that the structure of the FNN is fixed. Figure 3 shows the flowchart of training method SOS, which is started by collecting, normalizing, and reading a dataset. Once a network has been structured for a particular application, including setting the desired number of neurons in each layer, it is ready for training.

The Feedforward Neural Networks Architecture

When implementing a neural network, it is necessary to determine the structure based on the number of layers and the number of neurons in the layers. The larger the number of hidden layers and nodes, the more complex the network will be. In this work, the number of input and output neurons in MLP network is problem-dependent and the number of hidden nodes is computed on the basis of Kolmogorov theorem [56]: Hidden = 2× Input + 1. When using SOS to optimize the weights and bias in network, the dimension of each organism is considered as D, shown in

$$D = (\text{Input} \times \text{Hidden}) + (\text{Hidden} \times \text{Output}) + \text{Hidden}_{\text{bias}} + \text{Output}_{\text{bias}}, \tag{9}$$

where Input, Hidden, and Output refer to the number of input, hidden, and output neurons of FNN, respectively. Also, $\text{Hidden}_{\text{bias}}$ and $\text{Output}_{\text{bias}}$ are the number of biases in hidden and output layers.

Fitness Function

In SOS, every organism is evaluated according to its status (fitness). This evaluation is done by passing the vector of weights and biases to FNNs; then the mean squared error (MSE) criterion is calculated based on the prediction of the neural network using the training dataset. Through continuous iterations, the optimal solution is finally achieved, which is regarded as the weights and biases of a neural network. The MSE criterion is given in (10) where y and $\hat{y}$ are the actual and the estimated values based on proposed model and R is the number of samples in the training dataset.

$$\text{MSE} = \frac{1}{R}\sum_{i=1}^{R}(y - \hat{y})^2 \qquad (10)$$

Encoding Strategy

According to [57], the weights and biases of FNNs for every agent in evolutionary algorithms can be encoded and represented in the form of vector, matrix, or binary. In this work, the vector encoding method is utilized. An example of this encoding strategy for FNN is provided as shown in Figure 2. During the initialization process, $X = (X_1, X_2, \ldots,)$ is set on behalf of the N organisms. Each organism $X_i = \{\text{iw, hw, hb, ob}\}$ (i = 1, 2, . . . , N) represents complete set of FNN weights and biases, which is converted into a single vector of real number.

Criteria for Evaluating Performance

Classification is used to understand the existing data and to predict how unseen data will behave. In other words, the objective of data classification is to classify the unseen data in different classes on the basis of studying the existing data. For the classification problem, in addition to MSE criterion, accuracy rate was used. This rate measures the ability of the classifier by producing accurate results which can be computed as follows:

$$\text{Accuracy} = \frac{\tilde{N}}{N}, \qquad (11)$$

where $\tilde{N}$ represents the number of correctly classified objects by the classifier and N is the number of objects in the dataset.

SIMULATION EXPERIMENTS

This section presents a comprehensive analysis to investigate the efficiency of the SOS algorithm for training FNNs. As shown in Table 1, eight datasets are selected from UCI machine learning repository [58] to evaluate the performance of SOS. And six metaheuristic algorithms, including BBO [43], CS [37], GA [23], GSA [27, 28], PSO [28], and MVO [33], are presented for a reliable comparison.

Table 1: Description of datasets.

Dataset	Attribute	Class	Training sample	Testing sample	Input	Hidden	Output
Blood	4	2	493	255	4	9	2
Balance Scale	4	3	412	213	4	9	3
Haberman's Survival	3	2	202	104	3	7	2
Liver Disorders	6	2	227	118	6	13	2
Seeds	7	3	139	71	7	15	3
Wine	13	3	117	61	13	27	3
Iris	4	3	99	51	4	9	3
Statlog (Heart)	13	2	178	92	13	27	2

Datasets Design

The Blood dataset contains 748 instances, which were selected randomly from the donor database of Blood Transfusion Service Center in Hsinchu City in Taiwan. As a binary classification problem, the output class variable of the dataset represents whether the person donated blood in a time period (1 stands for donating blood; 0 stands for not donating blood). And the input variables are Recency, months since last donation; Frequency, total number of donation; Monetary: total blood donated in c.c.; and Time, months since first donation [59].

The Balance Scale dataset is generated to model psychological experiments reported by Siegler [60]. This dataset contains 625 examples and each example is classified as having the balance scale tip to the right and tip to the left or being balanced. The attributes are the left weight, the left distance, the right weight, and the right distance. The correct way to find the class is the greater of (left distance * left weight) and (right distance * right weight). If they are equal, it is balanced.

Haberman's Survival dataset contains cases from a study that was conducted between 1958 and 1970 at the University of Chicago's Billings Hospital on the survival of patients who had undergone surgery for breast cancer. The dataset contains 306 cases which record two survival status patients with age of patient at time of operation, patient's year of operation, and number of positive axillary nodes detected.

The Liver Disorders dataset was donated by BUPA Medical Research Ltd to record the liver disorder status in terms of a binary label. The dataset includes values of 6 features measured for 345 male individuals. The first 5 features are all blood tests which are thought to be sensitive to liver disorders that might arise from excessive alcohol consumption. These features are Mean Corpuscular Volume (MCV), alkaline phosphatase (ALKPHOS), alanine aminotransferase (SGPT), aspartate aminotransferase (SGOT), and gamma-glutamyl transpeptidase (GAMMAGT). The sixth feature is the number of alcoholic beverage drinks per day (DRINKS).

The Seeds dataset consists of 210 patterns belonging to three different varieties of wheat: Kama, Rosa, and Canadian. From each species there are 70 observations for area A , perimeter P , compactness C ($C = 4 * pi * A/P^2$),, length of kernel, width of kernel, asymmetry coefficient, and length of kernel groove.

The Wine dataset contains 178 instances recording the results of a chemical analysis of wines grown in the same region in Italy but derived from three different cultivars. The analysis determined the quantities of 13 constituents found in each of the three types of wines.

The Iris dataset contains 3 species of 50 instances each, where each species refers to a type of Iris plant (setosa, versicolor, and virginica). One species is linearly separable from the other 2 and the latter are not linearly separable from each other. Each of the 3 species is classified by three attributes: sepal length, sepal width, petal length, and petal width in cm. This dataset was used by Fisher [61] in his initiation of the linear-discriminate-function technique.

The Statlog (Heart) dataset is a heart disease database containing 270 instances that consist of 13 attributes: age, sex, chest pain type (4 values), resting blood pressure, serum cholesterol in mg/dL, fasting blood sugar > 120 mg/dL, resting electrocardiographic results (values 0, 1, and 2), maximum heart rate achieved, exercise induced angina, oldpeak = ST depression induced by exercise relative to rest, the slope of the peak exercise ST segment, number of major vessels (0–3) colored by fluoroscopy, and thal: 3 = normal; 6 = fixed defect; 7 = reversible defect.

Experimental Setup

In this section, the experiments were done using a desktop computer with a 3.30 GHz Intel(R) Core(TM) i5 processor, 4 GB of memory. The entire algorithm was programmed in MATLAB R2012a. The mentioned datasets are partitioned into 66% for training and 34% for testing [33]. All experiments are executed for 20 different runs and each run includes 500 iterations. The population size is considered as 30 and other control parameters of the corresponding algorithms are given below:

In CS, the possibility of eggs being detected and thrown out of the nest is $pa= 0.25$.

In GA, crossover rate $P_c = 0.5$; mutate rate $P_c = 0.05$.

In PSO, the parameters are set to $C_1 = C_2 = 2$; weight factor w decreased linearly from 0.9 to 0.5.

In BBO, mutation probability $p_m = 0.1$, the value for both max immigration (I) and max emigration (E) is 1, and the habitat modification probability $p_h = 0.8$.

In MVO, exploitation accuracy is set to p=6, the min traveling distance rate is set to 0.2, and the max traveling distance rate is set to 1. In GSA, α is set to 20, the gravitational constant (G_0) is set to 1, and initial values of acceleration and mass are set to 0 for each particle.

All input features are mapped onto the interval of [-1, 1] for a small scale. Here, we apply min-max normalization to perform a linear transformation on the original data as given in (12), where v› is the normalized value of v in the range [min, max].

$$v' = 2 * \frac{v - \min}{\max - \min} - 1 \qquad (12)$$

Results and Discussion

To evaluate the performance of the proposed method SOS with other six algorithms, BBO, CS, GA, GSA, PSO, and MVO, experiments are conducted using the given datasets. In this work, all datasets have been partitioned into two sets: training set and testing set. The training set is used to train the network in order to achieve the optimal weights and bias. The testing set is applied on unseen data to test the generalization performance of metaheuristic algorithms on FNNs.

Table 2 shows the best values of mean squared error (MSE), the worst values of MSE, the mean of MSE, and the standard deviation for all training datasets. Inspecting the table of results, it can be seen that SOS performs best in datasets Seeds and Iris. For datasets Liver Disorders, Haberman's Survival, and Blood, the best values, worst values, and mean values are all in the same order of magnitudes. While the values of MSE are smaller in SOS than the other algorithms, which means SOS is the best choice as the training method on the three aforementioned datasets. Moreover, it is ranked second for the dataset Wine and shows very competitive results compared to BBO. In datasets Balance Scale and Statlog (Heart), the best values in results indicate that SOS provides very close performances compared to BBO and MVO. Also the three algorithms show improvements compared to the others.

Table 2: MSE results.

Dataset	Algorithm						
	SOS	MVO	GSA	PSO	BBO	CS	GA
				Blood			
Best	$2.95E-01$	$3.04E-01$	$3.10E-01$	$3.07E-01$	$3.00E-01$	$3.11E-01$	$3.30E-01$
Worst	$3.05E-01$	$3.07E-01$	$3.33E-01$	$3.14E-01$	$3.92E-01$	$3.21E-01$	$4.17E-01$
Mean	$3.01E-01$	$3.05E-01$	$3.23E-01$	$3.10E-01$	$3.18E-01$	$3.17E-01$	$3.78E-01$
Std.	$2.44E-03$	$7.09E-04$	$6.60E-03$	$2.47E-03$	$2.84E-02$	$2.97E-03$	$2.85E-02$
				Balance Scale			
Best	$7.00E-02$	$8.03E-02$	$1.37E-01$	$1.40E-01$	$8.52E-02$	$1.70E-01$	$2.97E-01$
Worst	$1.29E-01$	$1.04E-01$	$1.66E-01$	$1.88E-01$	$1.24E-01$	$2.14E-01$	$8.01E-01$
Mean	$1.05E-01$	$8.66E-02$	$1.52E-01$	$1.72E-01$	$1.02E-01$	$1.89E-01$	$4.65E-01$
Std.	$1.33E-02$	$6.06E-03$	$9.46E-03$	$1.32E-02$	$9.45E-03$	$1.18E-02$	$1.16E-01$
				Haberman's Survival			
Best	$2.95E-01$	$3.21E-01$	$3.64E-01$	$3.43E-01$	$3.13E-01$	$3.61E-01$	$3.71E-01$
Worst	$3.37E-01$	$3.48E-01$	$4.07E-01$	$3.75E-01$	$3.51E-01$	$3.79E-01$	$4.77E-01$
Mean	$3.18E-01$	$3.31E-01$	$3.82E-01$	$3.65E-01$	$3.31E-01$	$3.70E-01$	$4.27E-01$
Std.	$1.04E-02$	$6.63E-03$	$1.14E-02$	$7.84E-03$	$1.07E-02$	$5.83E-03$	$3.45E-02$
				Liver Disorders			
Best	$3.26E-01$	$3.33E-01$	$4.17E-01$	$3.96E-01$	$3.16E-01$	$4.14E-01$	$5.27E-01$
Worst	$3.86E-01$	$3.55E-01$	$4.71E-01$	$4.31E-01$	$5.87E-01$	$4.57E-01$	$6.73E-01$
Mean	$3.49E-01$	$3.41E-01$	$4.44E-01$	$4.11E-01$	$3.82E-01$	$4.37E-01$	$5.95E-01$
Std.	$1.65E-02$	$6.17E-03$	$1.32E-02$	$1.06E-02$	$5.47E-02$	$1.12E-02$	$4.30E-02$

				Seeds			
Best	$9.78E-04$	$1.44E-02$	$5.97E-02$	$4.49E-02$	$2.07E-03$	$7.19E-02$	$2.05E-01$
Worst	$2.26E-02$	$3.30E-02$	$8.87E-02$	$1.02E-01$	$3.32E-01$	$1.22E-01$	$7.14E-01$
Mean	$1.11E-02$	$2.21E-02$	$7.65E-02$	$8.17E-02$	$5.23E-02$	$9.80E-02$	$4.71E-01$
Std.	$5.36E-03$	$6.31E-03$	$8.82E-03$	$1.53E-02$	$9.63E-02$	$1.34E-02$	$1.20E-01$
				Wine			
Best	$6.35E-11$	$1.34E-06$	$5.87E-04$	$9.74E-03$	$5.62E-12$	$2.56E-02$	$5.60E-01$
Worst	$8.55E-03$	$2.91E-05$	$1.90E-02$	$8.03E-02$	$3.42E-01$	$1.62E-01$	$8.95E-01$
Mean	$4.40E-04$	$5.41E-06$	$3.21E-03$	$3.39E-02$	$5.17E-02$	$9.59E-02$	$7.03E-01$
Std.	$1.91E-03$	$6.23E-06$	$3.96E-03$	$1.55E-02$	$1.25E-01$	$3.48E-02$	$8.59E-02$

				Iris			
Best	$5.10E-08$	$2.45E-02$	$4.41E-02$	$4.71E-02$	$1.70E-02$	$1.01E-02$	$1.19E-01$
Worst	$2.67E-02$	$2.75E-02$	$2.31E-01$	$1.76E-01$	$4.65E-01$	$7.84E-02$	$6.51E-01$
Mean	$1.42E-02$	$2.58E-02$	$6.83E-02$	$1.09E-01$	$6.91E-02$	$5.32E-02$	$3.66E-01$
Std.	$8.80E-03$	$8.40E-04$	$4.07E-02$	$3.55E-02$	$1.11E-01$	$1.93E-02$	$1.70E-01$
				Statlog (Heart)			
Best	$8.98E-02$	$5.03E-02$	$1.35E-01$	$1.72E-01$	$8.03E-02$	$2.31E-01$	$4.52E-01$
Worst	$1.26E-01$	$8.13E-02$	$1.92E-01$	$2.20E-01$	$1.65E-01$	$3.09E-01$	$6.97E-01$
Mean	$1.09E-01$	$6.48E-02$	$1.62E-01$	$1.93E-01$	$1.27E-01$	$2.61E-01$	$5.53E-01$
Std.	$1.07E-02$	$9.11E-03$	$1.47E-02$	$1.32E-02$	$2.37E-02$	$1.86E-02$	$7.20E-02$

Convergence curves for all metaheuristic algorithms are shown in Figures 4, 6, 8, 10, 12, 14, 16, and 18. The convergence curves show the average of 20 independent runs over the course of 500 iterations. The figures show that SOS has the fastest convergence speed for training all the given datasets. Figures 5, 7, 9, 11, 13, 15, 17, and 19 show the boxplots relative to 20 runs of SOS, BBO, GA, MVO, PSO, GSA, and CS. The boxplots, which are used to analyze the variability in getting MSE values, indicate that SOS has greater value and less height than those of SOS, GA, CS, PSO, and GSA and achieves the similar results to MVO and BBO.

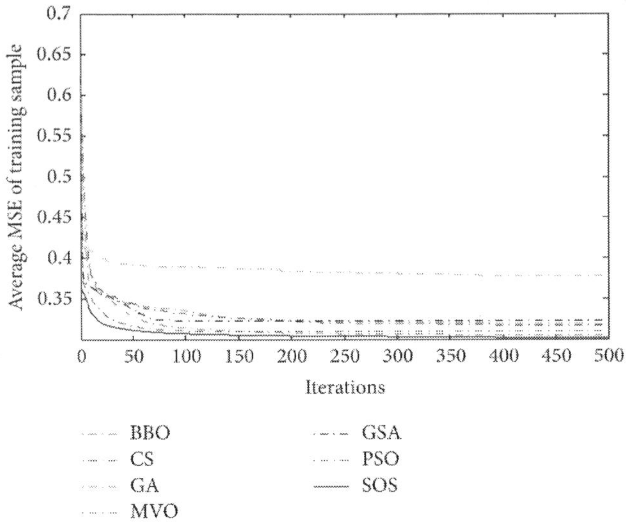

Figure 4: The convergence curves of algorithms (Blood).

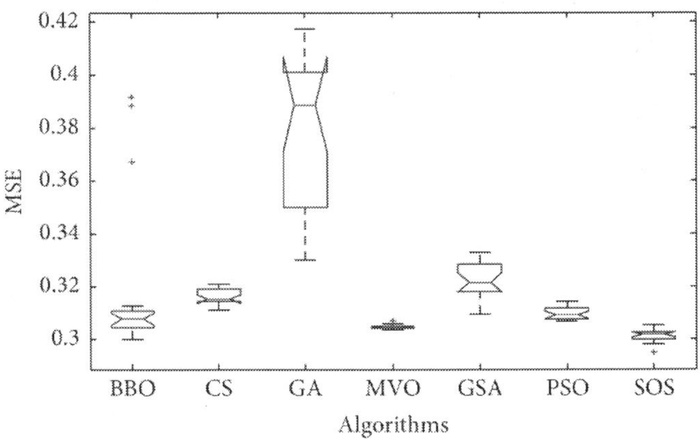

Figure 5: The ANOVA test of algorithms for training Blood.

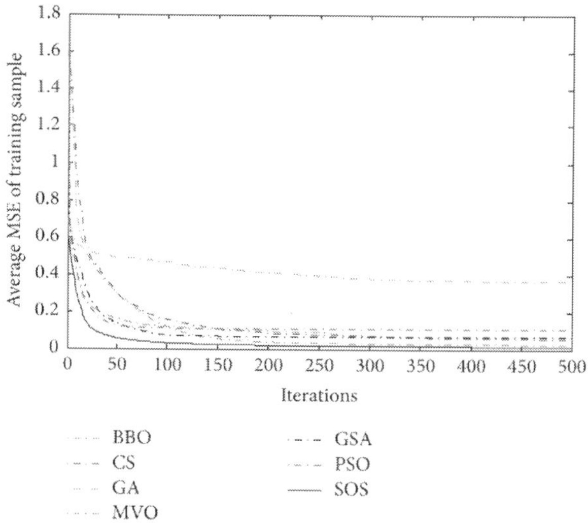

Figure 6: The convergence curves of algorithms (Iris).

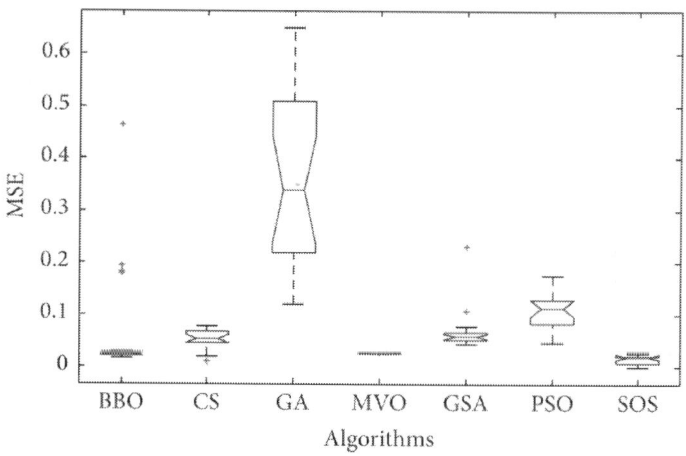

Figure 7: The ANOVA test of algorithms for training Iris.

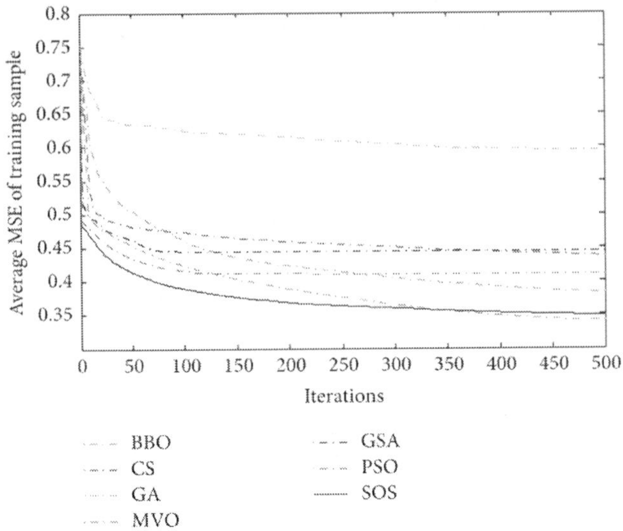

Figure 8: The convergence curves of algorithms (Liver Disorders).

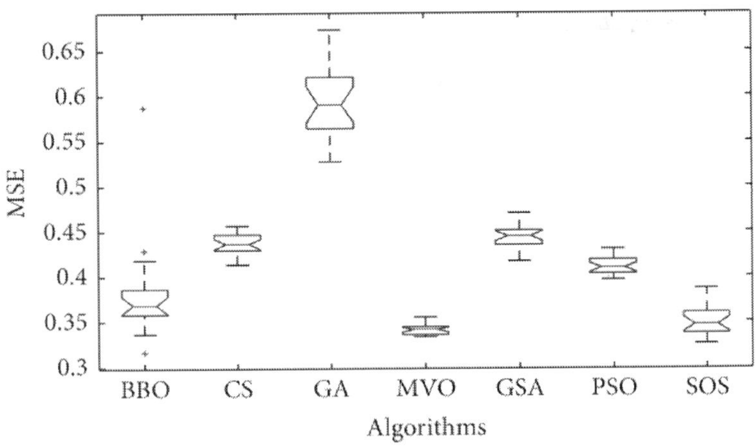

Figure 9: The ANOVA test of algorithms for training Liver Disorders.

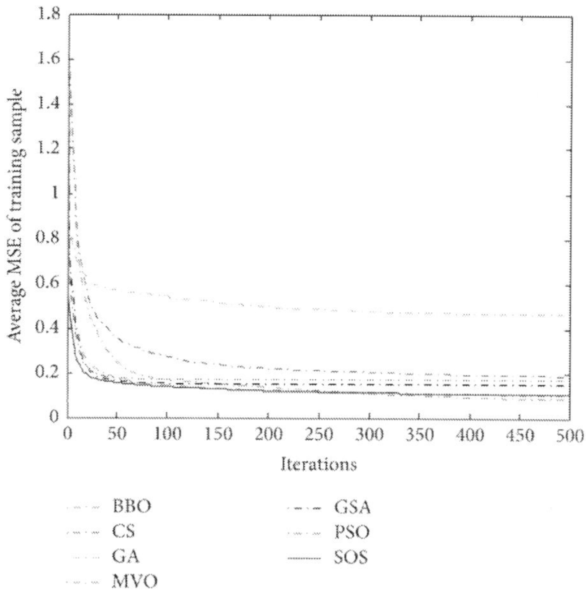

Figure 10: The convergence curves of algorithms (Balance Scale).

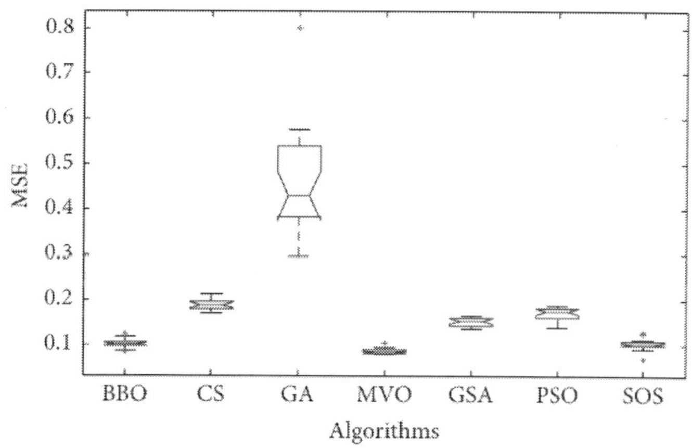

Figure 11: The ANOVA test of algorithms for training Balance Scale.

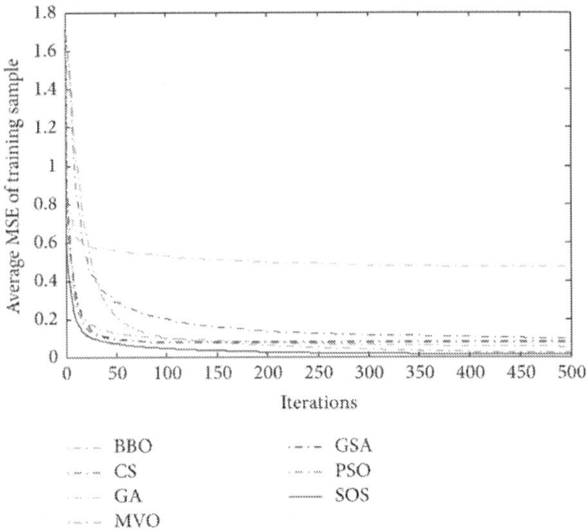

Figure 12: The convergence curves of algorithms (Seeds).

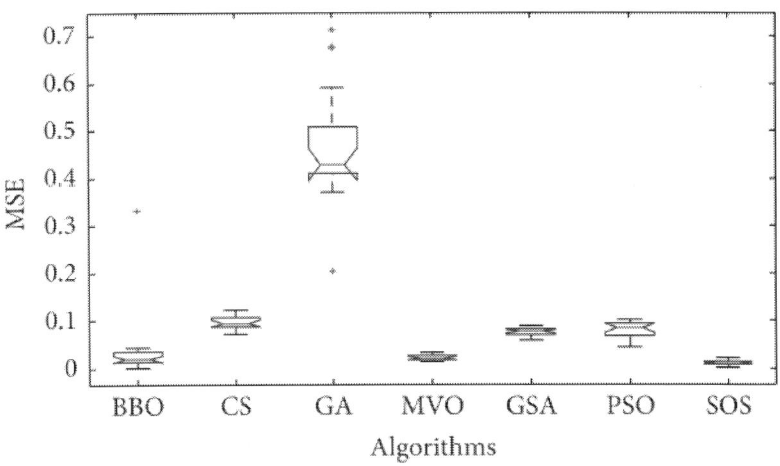

Figure 13: The ANOVA test of algorithms for training Seeds.

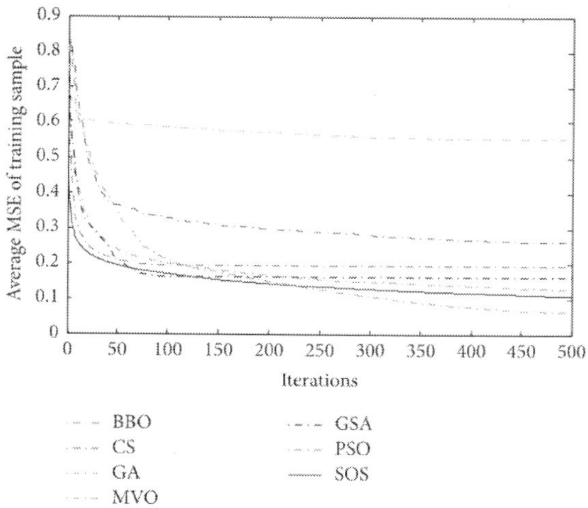

Figure 14: The convergence curves of algorithms (Statlog (Heart)).

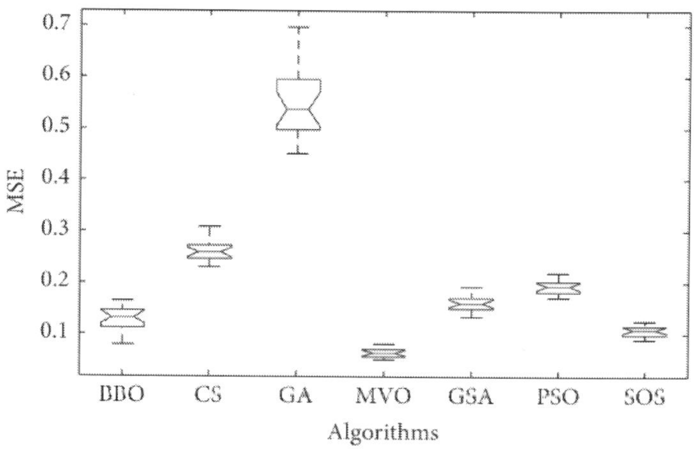

Figure 15: The ANOVA test of algorithms for training Statlog (Heart).

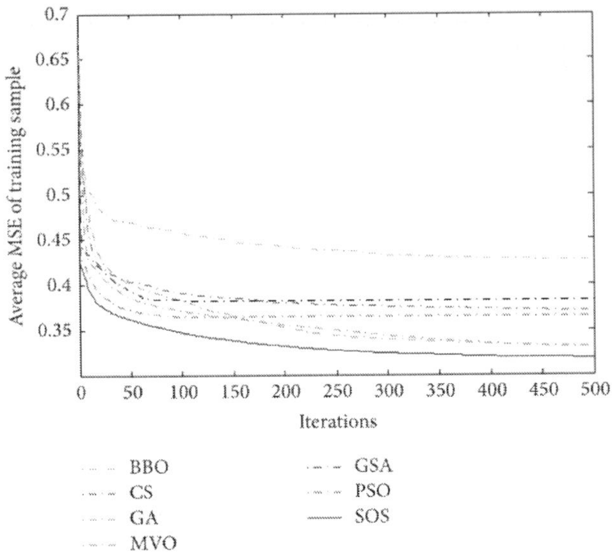

Figure 16: The convergence curves of algorithms (Haberman's Survival).

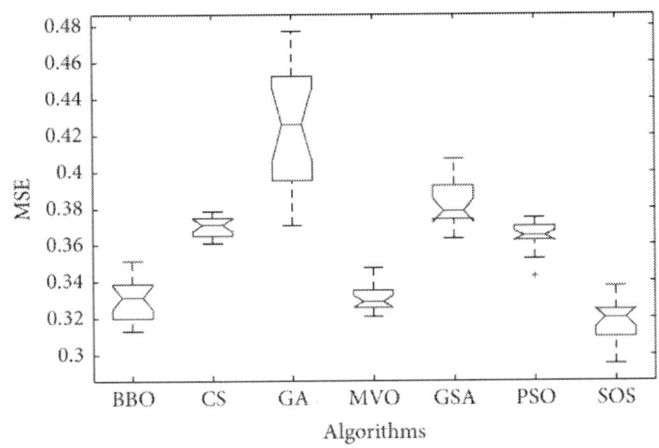

Figure 17: The ANOVA test of algorithms for training Haberman's Survival.

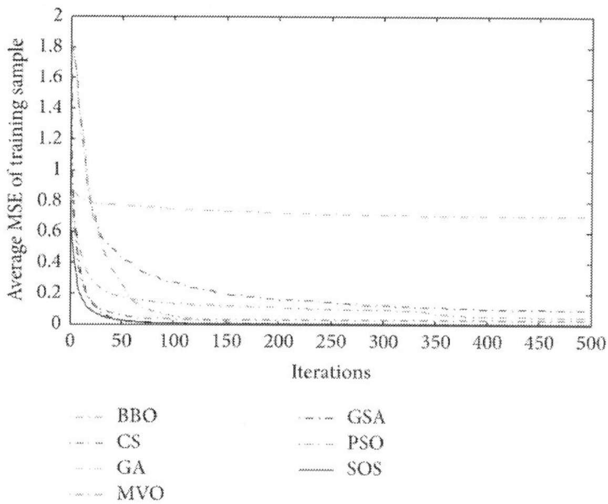

Figure 18: The convergence curves of algorithms (Wine).

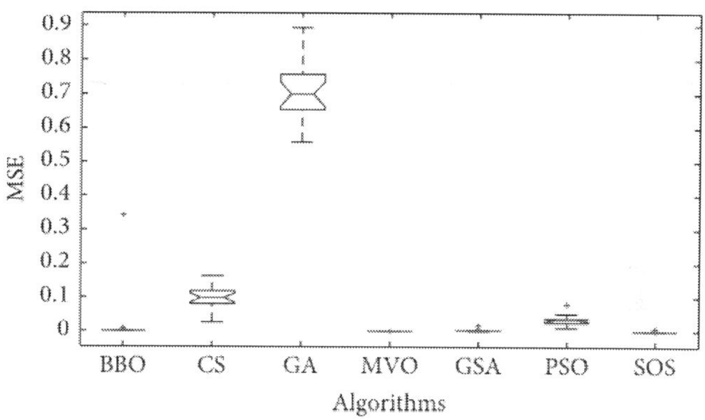

Figure 19: The ANOVA test of algorithms for training Wine.

Through 20 independent runs on the training datasets, the optimal weights and biases are achieved and then used to test the classification accuracy on the testing datasets. As depicted in Table 3, the rank is in terms of the best values in each dataset and SOS provides the best performances on testing datasets: Blood, Seeds, and Iris. For dataset Wine, the classification accuracy of SOS is 98.3607% which indicates that only one example in testing dataset cannot be classified correctly. It is noticeable that, though MVO has the highest classification accuracy in datasets Balance Scale, Haberman's Survival, Liver Disorders, and

Statlog (Heart), SOS also performs well in classification. However, the accuracy shown in GA is the lowest among the tested algorithms.

Table 3: Accuracy results.

Dataset	Algorithm						
	SOS	MVO	GSA	PSO	BBO	CS	GA
				Blood			
Best	82.7451	81.1765	78.0392	80.3922	81.1765	78.8235	76.8627
Worst	77.6471	80.0000	33.3333	77.6471	72.5490	74.5098	67.0588
Mean	79.8039	80.7451	74.1765	79.2157	77.5294	76.9608	72.7843
Rank	1	2	6	4	2	5	7
				Balance Scale			
Best	92.0188	92.9577	91.0798	89.2019	91.5493	90.6103	80.2817
Worst	86.8545	89.2019	85.9155	83.5681	88.2629	82.1596	38.0282
Mean	90.0235	91.4319	87.7465	86.7136	90.1643	86.3146	59.7653
Rank	2	1	4	6	3	5	7
				Haberman's Survival			
Best	81.7308	82.6923	81.7308	82.6923	82.6923	81.7308	78.8462
Worst	71.1538	75.9615	74.0385	76.9231	69.2308	74.0385	65.3846
Mean	76.0577	79.5673	79.1827	80.4808	77.0192	78.2692	74.5673
Rank	4	1	4	1	1	6	7
				Liver Disorders			
Best	75.4237	76.2712	64.4068	72.0339	72.8814	67.7966	55.0847
Worst	66.1017	72.0339	6.7797	59.3220	45.7627	47.4576	27.1186
Mean	71.0593	74.1949	48.3051	66.4831	65.1271	56.1441	43.0508
Rank	2	1	6	4	3	5	7

				Seeds			
Best	95.7746	95.7746	94.3662	92.9577	94.3662	91.5493	67.6056
Worst	87.3239	90.1408	85.9155	78.8732	61.9718	77.4648	28.1690
Mean	91.3380	93.4507	90.3521	87.4648	87.3239	82.2535	51.1268
Rank	1	1	3	5	3	6	7
				Wine			
Best	98.3607	98.3607	100.0000	100.0000	100.0000	91.8033	49.1803
Worst	91.8033	96.7213	91.8033	88.5246	59.0164	78.6885	18.0328
Mean	95.4918	97.9508	96.1475	96.0656	90.4918	83.3607	32.6230
Rank	4	4	1	1	1	6	7
				Iris			
Best	98.0392	98.0392	98.0392	98.0392	98.0392	98.0392	98.0392
Worst	64.7059	98.0392	33.3333	52.9412	29.4118	52.9412	7.8431
Mean	92.0588	98.0392	93.7255	91.4706	82.9412	85.0000	56.0784
Rank	1	1	1	1	1	1	1
				Statlog (Heart)			
Best	85.8696	88.0435	86.9565	86.9565	80.4348	83.6957	70.6522
Worst	77.1739	77.1739	33.6957	77.1739	66.3043	68.4783	39.1304
Mean	82.2283	82.6087	79.1848	82.8804	75.5435	77.4457	50.5435
Rank	4	1	2	2	6	5	7

This comprehensive comparative study shows that the SOS algorithm is superior among the compared trainers in this paper. It is a challenge for training FNN due to the large number of local solutions in solving this problem. On account of being simpler and more robust than competing algorithms, SOS performs well in most of the datasets, which shows how flexible this algorithm is for solving problems with diverse search space. Further, in order to determine whether the results achieved by the algorithms are statistically different from each other, a nonparametric statistical significance proof known as Wilcoxon's rank sum test for equal medians [62, 63] was conducted between the results obtained by the algorithms, SOS versus CS, SOS versus PSO, SOS versus GA, SOS versus MVO, SOS versus GSA, and SOS versus BBO. In order to draw a statistically meaningful conclusion, tests are performed on the optimal fitness for training datasets and P values are computed as shown in Table 4. Rank sum tests the null hypothesis that the two datasets are samples from continuous distributions with equal medians, against the alternative that they are not. Almost all values reported in Table 4 are less than 0.05 (5% significant level) which is strong evidence against the null hypothesis. Therefore, such evidence indicates that SOS results are statistically significant and that it has not occurred by coincidence (i.e., due to common noise contained in the process).

Table 4: P values produced by Wilcoxon's rank sum test for equal medians.

Dataset	SOS versus					
	CS	PSO	GA	MVO	GSA	BBO
Blood	$6.80E-08$	$6.80E-08$	$6.80E-08$	$3.07E-06$	$6.80E-08$	$4.68E-05$
Balance Scale	$6.80E-08$	$6.80E-08$	$6.80E-08$	$9.75E-06$	$6.80E-08$	$2.29E-01$
Haberman's Survival	$6.80E-08$	$6.80E-08$	$6.80E-08$	$1.79E-04$	$6.80E-08$	$2.56E-03$
Liver Disorders	$6.80E-08$	$6.80E-08$	$6.80E-08$	$1.26E-01$	$6.80E-08$	$1.35E-03$
Seeds	$6.80E-08$	$6.80E-08$	$6.80E-08$	$3.99E-06$	$6.80E-08$	$1.63E-03$
Wine	$6.78E-08$	$6.80E-08$	$6.80E-08$	$1.48E-03$	$1.05E-06$	$5.98E-01$
Iris	$2.96E-06$	$6.47E-08$	$6.47E-08$	$7.62E-07$	$6.47E-08$	$5.73E-05$
Statlog (Heart)	$6.80E-08$	$6.80E-08$	$6.80E-08$	$6.80E-08$	$6.80E-08$	$6.04E-03$

Analysis of the Results

Statistically speaking, the SOS algorithm provides superior local avoidance and the high classification accuracy in training FNNs. According to the mathematical formulation of the SOS algorithm, the first two interaction phases are devoted to exploration of the search space. This promotes exploration of the search space that leads to finding the optimal weights and biases. For the exploitation phase, the third interaction phase of SOS algorithm is helpful for resolving local optima stagnation. The results of this work show that although metaheuristic optimizations have high exploration, the problem of training an FNN needs high local optima avoidance during the whole optimization process. The results prove that the SOS is very effective in training FNNs.

It is worth discussing the poor performance of GA in this subsection. The rate of crossover and mutation are two specific tuning parameters in GA, dependent on the empirical value for particular problems. This is the reason why GA failed to provide good results for all the datasets. In the contrast, SOS uses only the two parameters of maximum evaluation number and population size, so it avoids the risk of compromised performance due to improper parameter tuning and enhances performance stability. Easy to fall into local optimal and low efficiency in the latter of search period are the other two reasons for the poor performance of GA. Another finding in the results is the good performances of BBO and MVO which are benefit from the mechanism for significant abrupt movements in the search space.

The reason for the high classification rate provided by SOS is that this algorithm is equipped with adaptive three phases to smoothly balance exploration and exploitation. The first two phases are devoted to exploration and the rest to exploitation. And the three phases are simple to operate with only simple mathematical operations to code. In addition, SOS uses greedy selection at the end of each phase to select whether to retain the old or modified solution. Consequently, there are always guiding search agents to the most promising regions of the search space.

CONCLUSIONS

In this paper, the recently proposed SOS algorithm was employed for the first time as a FNN trainer. The high level of exploration and exploitation of this algorithm were the motivation for this study. The problem of training a FNN was first formulated for the SOS algorithm. This algorithm was then employed to optimize the weights and biases of FNNs so as to get high classification accuracy. The obtained results of eight datasets with different characteristic show that the proposed approach is efficient to train FNNs compared to other training methods that have been used in the literatures: CS, PSO, GA, MVO, GSA, and BBO. The results of MSE over 20 runs show that the proposed approach performs best in terms of convergence rate and is robust since the variances are relatively small. Furthermore, by comparing the classification accuracy of the testing datasets, using the optimal weights and biases, SOS has advantage over the other algorithms employed. In addition, the significance of the results is statistically confirmed by using Wilcoxon's rank sum test, which demonstrates that the results have not occurred by coincidence. It can be concluded that SOS is suitable for being used as a training method for FNNs.

For future work, the SOS algorithm will be extended to find the optimal number of layers, hidden nodes, and other structural parameters of FNNs. More elaborate tests on higher dimensional problems and large number of datasets will be done. Other types of neural networks such as radial basis function (RBF) neural network are worth further research.

Competing Interests

The authors declare that there is no conflict of interests regarding the publication of this paper.

Acknowledgments

This work is supported by National Science Foundation of China under Grants no. 61463007 and 61563008 and the Guangxi Natural Science Foundation under Grant no. 2016GXNSFAA380264.

REFERENCES

1. J. A. Anderson, An Introduction to Neural Networks, MIT Press, 1995.
2. G. P. Zhang and M. Qi, "Neural network forecasting for seasonal and trend time series," European Journal of Operational Research, vol. 160, no. 2, pp. 501–514, 2005. ·
3. C.-J. Lin, C.-H. Chen, and C.-Y. Lee, "A self-adaptive quantum radial basis function network for classification applications," in Proceedings of the IEEE International Joint Conference on Neural Networks, vol. 4, pp. 3263–3268, IEEE, Budapest, Hungary, July 2004. ·
4. X.-F. Wang and D.-S. Huang, "A novel density-based clustering framework by using level set method," IEEE Transactions on Knowledge and Data Engineering, vol. 21, no. 11, pp. 1515–1531, 2009.
5. N. A. Mat Isa and W. M. F. W. Mamat, "Clustered-Hybrid Multilayer Perceptron network for pattern recognition application," Applied Soft Computing Journal, vol. 11, no. 1, pp. 1457–1466, 2011.
6. D. S. Huang, Systematic Theory of Neural Networks for Pattern Recognition, Publishing House of Electronic Industry of China,

1996 (Chinese).

7. Z.-Q. Zhao, D.-S. Huang, and B.-Y. Sun, "Human face recognition based on multi-features using neural networks committee," Pattern Recognition Letters, vol. 25, no. 12, pp. 1351–1358, 2004.
8. O. Nelles, "Nonlinear system identification: from classical approaches to neural networks and fuzzy models," Applied Therapeutics, vol. 6, no. 7, pp. 21–717, 2001.
9. B. Malakooti and Y. Q. Zhou, "Approximating polynomial functions by feedforward artificial neural networks: capacity analysis and design," Applied Mathematics and Computation, vol. 90, no. 1, pp. 27–51, 1998.
10. A. Cochocki and R. Unbehauen, Neural Networks for Optimization and Signal Processing, John Wiley & Sons, New York, NY, USA, 1993.
11. X.-F. Wang, D.-S. Huang, and H. Xu, "An efficient local Chan-Vese model for image segmentation," Pattern Recognition, vol. 43, no. 3, pp. 603–618, 2010.
12. D.-S. Huang, "A constructive approach for finding arbitrary roots of polynomials by neural networks," IEEE Transactions on Neural Networks, vol. 15, no. 2, pp. 477–491, 2004.
13. G. Bebis and M. Georgiopoulos, "Feed-forward neural networks," IEEE Potentials, vol. 13, no. 4, pp. 27–31, 1994.
14. T. Kohonen, "The self-organizing map," Neurocomputing, vol. 21, no. 1–3, pp. 1–6, 1998.
15. J. Park and I. W. Sandberg, "Approximation and radial-basis-function networks," Neural Computation, vol. 3, no. 2, pp. 246–257, 1993.
16. D.-S. Huang, "Radial basis probabilistic neural networks: model and application," International Journal of Pattern Recognition and Artificial Intelligence, vol. 13, no. 7, pp. 1083–1101, 1999.
17. D.-S. Huang and J.-X. Du, "A constructive hybrid structure optimization methodology for radial basis probabilistic neural networks," IEEE Transactions on Neural Networks, vol. 19, no. 12, pp. 2099–2115, 2008.
18. G. Dorffner, "Neural networks for time series processing," Neural

Network World, vol. 6, no. 1, pp. 447–468, 1996.

19. S. Ghosh-Dastidar and H. Adeli, "Spiking neural networks," International Journal of Neural Systems, vol. 19, no. 4, pp. 295–308, 2009.
20. D. E. Rumelhart, E. G. Hinton, and R. J. Williams, "Learning representations by back-propagating errors," in Neurocomputing: Foundations of Research, vol. 5, no. 3, pp. 696–699, MIT Press, Cambridge, Mass, USA, 1988.
21. N. Zhang, "An online gradient method with momentum for two-layer feedforward neural networks," Applied Mathematics and Computation, vol. 212, no. 2, pp. 488–498, 2009. ·
22. J. H. Holland, Adaptation in Natural and Artificial Systems, MIT Press, Cambridge, Mass, USA, 1992.
23. D. J. Montana and L. Davis, "Training feedforward neural networks using genetic algorithms," in Proceedings of the International Joint Conferences on Artificial Intelligence (IJCAI '89), vol. 89, pp. 762–767, Detroit, Mich, USA, 1989.
24. C. R. Hwang, "Simulated annealing: theory and applications," Acta Applicandae Mathematicae, vol. 12, no. 1, pp. 108–111, 1988.
25. D. Shaw and W. Kinsner, "Chaotic simulated annealing in multilayer feedforward networks," in Proceedings of the 1996 Canadian Conference on Electrical and Computer Engineering (CCECE'96), part 1 (of 2), pp. 265–269, University of Calgary, Alberta, Canada, May 1996.
26. J.-R. Zhang, J. Zhang, T.-M. Lok, and M. R. Lyu, "A hybrid particle swarm optimization-back-propagation algorithm for feedforward neural network training," Applied Mathematics & Computation, vol. 185, no. 2, pp. 1026–1037, 2007.
27. E. Rashedi, H. Nezamabadi-Pour, and S. Saryazdi, "GSA: a gravitational search algorithm," Information Sciences, vol. 179, no. 13, pp. 2232–2248, 2009.
28. S. A. Mirjalili, S. Z. M. Hashim, and H. M. Sardroudi, "Training feedforward neural networks using hybrid particle swarm optimization and gravitational search algorithm," Applied Mathematics and Computation, vol. 218, no. 22, pp. 11125–11137, 2012. ·

29. Z. Beheshti, S. M. H. Shamsuddin, E. Beheshti, and S. S. Yuhaniz, "Enhancement of artificial neural network learning using centripetal accelerated particle swarm optimization for medical diseases diagnosis," Soft Computing, vol. 18, no. 11, pp. 2253–2270, 2014.
30. L. A. M. Pereira, D. Rodrigues, P. B. Ribeiro, J. P. Papa, and S. A. T. Weber, "Social-spider optimization-based artificial neural networks training and its applications for Parkinson's disease identification," in Proceedings of the 27th IEEE International Symposium on Computer-Based Medical Systems (CBMS '14), pp. 14–17, IEEE, New York, NY, USA, May 2014.
31. E. Uzlu, M. Kankal, A. Akpınar, and T. Dede, "Estimates of energy consumption in Turkey using neural networks with the teaching-learning-based optimization algorithm," Energy, vol. 75, pp. 295–303, 2014.
32. P. A. Kowalski and S. Łukasik, "Training neural networks with Krill Herd algorithm," Neural Processing Letters, vol. 44, no. 1, pp. 5–17, 2016.
33. H. Faris, I. Aljarah, and S. Mirjalili, "Training feedforward neural networks using multi-verse optimizer for binary classification problems," Applied Intelligence, vol. 45, no. 2, pp. 322–332, 2016.
34. J. Nayak, B. Naik, and H. S. Behera, "A novel nature inspired firefly algorithm with higher order neural network: performance analysis," Engineering Science and Technology, vol. 19, no. 1, pp. 197–211, 2016. ·
35. C. Blum and K. Socha, "Training feed-forward neural networks with ant colony optimization: an application to pattern classification," in Proceedings of the 5th International Conference on Hybrid Intelligent Systems (HIS ‹05), pp. 233–238, IEEE Computer Society, Rio de Janeiro, Brazil, 2005.
36. K. Socha and C. Blum, "An ant colony optimization algorithm for continuous optimization: application to feed-forward neural network training," Neural Computing and Applications, vol. 16, no. 3, pp. 235–247, 2007.
37. E. Valian, S. Mohanna, and S. Tavakoli, "Improved cuckoo search algorithm for feedforward neural network training," International Journal of Artificial Intelligence & Applications, vol. 2, no. 3, pp.

36–43, 2011.

38. D. Karaboga, B. Akay, and C. Ozturk, "Artificial Bee Colony (ABC) optimization algorithm for training feed-forward neural networks," in Proceedings of the International Conference on Modeling Decisions for Artificial Intelligence (MDAI ‹07), pp. 318–329, Springer, Kitakyushu, Japan, 2007.
39. C. Ozturk and D. Karaboga, "Hybrid Artificial Bee Colony algorithm for neural network training," in Proceedings of the IEEE Congress of Evolutionary Computation (CEC ‹11), pp. 84–88, IEEE, New Orleans, La, USA, June 2011.
40. L. A. M. Pereira, L. C. S. Afonso, J. P. Papa et al., "Multilayer perceptron neural networks training through charged system search and its application for non-technical losses detection," in Proceedings of the 2nd IEEE Latin American Conference on Innovative Smart Grid Technologies (ISGT LA ‹13), pp. 1–6, IEEE, São Paulo, Brazil, April 2013.
41. S. Mirjalili, "How effective is the Grey Wolf optimizer in training multi-layer perceptrons," Applied Intelligence, vol. 43, no. 1, pp. 150–161, 2015.
42. P. Moallem and N. Razmjooy, "A multi layer perceptron neural network trained by invasive weed optimization for potato color image segmentation," Trends in Applied Sciences Research, vol. 7, no. 6, pp. 445–455, 2012.
43. S. Mirjalili, S. M. Mirjalili, and A. Lewis, "Let a biogeography-based optimizer train your multi-layer perceptron," Information Sciences. An International Journal, vol. 269, pp. 188–209, 2014. ·
44. M.-Y. Cheng and D. Prayogo, "Symbiotic organisms search: a new metaheuristic optimization algorithm," Computers & Structures, vol. 139, pp. 98–112, 2014.
45. M. Cheng, D. Prayogo, and D. Tran, "Optimizing multiple-resources leveling in multiple projects using discrete symbiotic organisms search," Journal of Computing in Civil Engineering, vol. 30, no. 3, 2016.
46. R. Eki, F. Vincent Y, S. Budi, and A. A. N. Perwira Redi, "Symbiotic Organism Search (SOS) for solving the capacitated vehicle routing problem," World Academy of Science, Engineering and

Technology, International Journal of Mechanical, Aerospace, Industrial, Mechatronic and Manufacturing Engineering, vol. 9, no. 5, pp. 807–811, 2015.

47. D. Prasad and V. Mukherjee, "A novel symbiotic organisms search algorithm for optimal power flow of power system with FACTS devices," Engineering Science & Technology, vol. 19, no. 1, pp. 79–89, 2016. ·

48. M. Abdullahi, M. A. Ngadi, and S. M. Abdulhamid, "Symbiotic organism Search optimization based task scheduling in cloud computing environment," Future Generation Computer Systems, vol. 56, pp. 640–650, 2016.

49. S. Verma, S. Saha, and V. Mukherjee, "A novel symbiotic organisms search algorithm for congestion management in deregulated environment," Journal of Experimental and Theoretical Artificial Intelligence, pp. 1–21, 2015.

50. D.-H. Tran, M.-Y. Cheng, and D. Prayogo, "A novel Multiple Objective Symbiotic Organisms Search (MOSOS) for time-cost-labor utilization tradeoff problem," Knowledge-Based Systems, vol. 94, pp. 132–145, 2016.

51. V. F. Yu, A. A. Redi, C. Yang, E. Ruskartina, and B. Santosa, "Symbiotic organisms search and two solution representations for solving the capacitated vehicle routing problem," Applied Soft Computing, 2016.

52. A. Panda and S. Pani, "A Symbiotic Organisms Search algorithm with adaptive penalty function to solve multi-objective constrained optimization problems," Applied Soft Computing, vol. 46, pp. 344–360, 2016. ·

53. S. Banerjee and S. Chattopadhyay, "Power optimization of three dimensional turbo code using a novel modified symbiotic organism search (MSOS) algorithm," Wireless Personal Communications, pp. 1–28, 2016.

54. B. Das, V. Mukherjee, and D. Das, "DG placement in radial distribution network by symbiotic organisms search algorithm for real power loss minimization," Applied Soft Computing, vol. 49, pp. 920–936, 2016. ·

55. M. K. Dosoglu, U. Guvenc, S. Duman, Y. Sonmez, and H. T.

Kahraman, “Symbiotic organisms search optimization algorithm for economic/emission dispatch problem in power systems,” Neural Computing and Applications, 2016.

56. R. Hecht-Nielsen, “Kolmogorov›s mapping neural network existence theorem,” in Proceedings of the IEEE 1st International Conference on Neural Networks, vol. 3, pp. 11–13, IEEE Press, San Diego, Calif, USA, 1987.
57. J.-R. Zhang, J. Zhang, T.-M. Lok, and M. R. Lyu, “A hybrid particle swarm optimization-back-propagation algorithm for feedforward neural network training,” Applied Mathematics and Computation, vol. 185, no. 2, pp. 1026–1037, 2007.
58. C. L. Blake and C. J. Merz, UCI Repository of Machine Learning Databases, http://archive.ics.uci.edu/ml/datasets.html.
59. I.-C. Yeh, K.-J. Yang, and T.-M. Ting, “Knowledge discovery on RFM model using Bernoulli sequence,” Expert Systems with Applications, vol. 36, no. 3, pp. 5866–5871, 2009.
60. R. S. Siegler, “Three aspects of cognitive development,” Cognitive Psychology, vol. 8, no. 4, pp. 481–520, 1976.
61. R. A. Fisher, “Has Mendel›s work been rediscovered,” Annals of Science, vol. 1, no. 2, pp. 115–137, 1936.
62. F. Wilcoxon, “Individual comparisons by ranking methods,” Biometrics Bulletin, vol. 1, no. 6, pp. 80–83, 1945.
63. J. Derrac, S. García, D. Molina, and F. Herrera, “A practical tutorial on the use of nonparametric statistical tests as a methodology for comparing evolutionary and swarm intelligence algorithms,” Swarm and Evolutionary Computation, vol. 1, no. 1, pp. 3–18, 2011. ·

CITATIONS

CHAPTER 1

Claudio Messori, Intelligence vs. Artificial Intelligence: The King Is Naked, 10.4236/oalib.1103115.

CHAPTER 2

Tielin Zhang, Yi Zeng, Bo Xu; Biological Neural Network Structure and Spike Activity Prediction Based on Multi-Neuron Spike Train Data; DOI: 10.4236/ijis.2015.52010 .

CHAPTER 3

Necmettin Firat Ozkan, Emin Kahya; An Experiment in Use of Brain Computer Interfaces for Cognitive Researches; DOI: 10.4236/ijis.2015.52008.

CHAPTER 4

Chenxia Gu, Shaotong Wang, Hao Yu; A Chessboard Model of Human Brain and An Application on Memory Capacity; DOI:10.4236/jamp.2016.42042.

CHAPTER 5

Narusci S. Bastos, Diana F. Adamatti, and Cleo Z. Billa, "Discovering Patterns in Brain Signals Using Decision Trees," Computational Intelligence and Neuroscience, vol. 2016, Article ID 6391807, 10 pages, 2016. doi:10.1155/2016/6391807.

CHAPTER 6

Romain Grandchamp and Arnaud Delorme, The Brainarium: An Interactive Immersive Tool for Brain Education, Art, and Neurotherapy, doi.org/10.1155/2016/4204385

CHAPTER 7

Terence Lovat, Daniel Fleming, Creativity as Central to Critical Reasoning and the Facilitative Role of Moral Education: Utilizing Insights from Neuroscience; DOI:10.4236/ce.2015.611108.

CHAPTER 8

Hong Gi Yeom, June Sic Kim, and Chun Kee Chung, "Macroscopic Neural Oscillation during Skilled Reaching Movements in Humans," Computational Intelligence and Neuroscience, vol. 2016, Article ID 2714052, 8 pages, 2016. doi:10.1155/2016/2714052.

CHAPTER 9

Kai Li, Xenophon Papademetris, and Don M. Tucker, "BrainK for Structural Image Processing: Creating Electrical Models of the Human Head," Computational Intelligence and Neuroscience, vol. 2016, Article ID 1349851, 25 pages, 2016. doi:10.1155/2016/1349851.

CHAPTER 10

Jianping Li, Quantum-Inspired Neural Networks with Application; DOI: 10.4236/ojapps.2015.56024.

CHAPTER 11

Haizhou Wu, Yongquan Zhou, Qifang Luo, and Mohamed Abdel Basset; Training Feedforward Neural Networks Using Symbiotic Organisms Search Algorithm; http://dx.doi.org/10.1155/2016/9063065.

INDEX

D

E

F

G

H

I

J

K

L

M

N

O

P

Q

R

S

T

U

V

W